New Dramaturgies

Series Editors
Cathy Turner
Drama Department
University of Exeter Drama Department
Exeter, UK

Synne Behrndt
Royal Institute of Technology
Stockholm, Sweden

This series explores new dramaturgies within contemporary performance practice and deploys dramaturgical thinking as a productive analytical and practical approach to both performance analysis and performance-making. Designed to inspire students, scholars and practitioners, the series extends the understanding of the complex contexts of dramaturgy and embraces its diversity and scope.

More information about this series at
http://www.palgrave.com/gp/series/14742

Lisa Woynarski

Ecodramaturgies

Theatre, Performance and Climate Change

palgrave
macmillan

Lisa Woynarski
University of Reading
Reading, UK

New Dramaturgies
ISBN 978-3-030-55852-9 ISBN 978-3-030-55853-6 (eBook)
https://doi.org/10.1007/978-3-030-55853-6

Cover illustration: *Trans-Plantable Living*, Cardiff, Wales, 2013. Photo: Nigel Pugh

This Palgrave Macmillan imprint is published by the registered company Springer Nature Switzerland AG.
The registered company address is: Gewerbestrasse 11, 6330 Cham, Switzerland

For Sally
with gratitude

Series Editors' Preface

The series New Dramaturgies is afforded by recent developments in the discussion of dramaturgy: a significant number of English-language publications now exist that offer a range of introductory approaches to the field, frequently by discussing the work of the dramaturg. Given the greater understanding this body of work enables, it is now possible to explore the subject and process of dramaturgy without centring on the explication of fundamental terms and the division of roles, but rather approaching it from a range of perspectives and in relation to emerging debates and performance forms. While at times this may include further enquiry into the dramaturg's specific role, we also hope that the series will make a significant contribution through the deployment of dramaturgical thinking as an approach to performance analysis and performance-making.

If dramaturgical practice entails the facilitation of practical decisions by way of interpretation and critical perspective, dramaturgical analysis concerns attention to detail in relation to a wider whole. Dramaturgy has been characterised as being about making connections, moving between elements, forming organic wholes which are continually in process; this also implies attention to audience and context. Dramaturgy, then, entails a discussion of composition in terms of process and event, rather than the self-contained and singular art work. We note that dramaturgy's historical association with literature, combined with its intrinsically holistic approach to the theatre event, enables movement and comparison across dramatic, postdramatic and other performance forms, without embedding divisions between them. It is also possible to expand the concept of dramaturgy to enable the discussion of performance in a wider, cultural sense. In this respect there are resonances with both sociology and performance studies.

Thus, while the series is partly concerned with dramaturgy as a professional and research field, it is equally a means to a discussion of contemporary performance, performance methodology and cultural context, through an address to the composition of action and event—or series of events. The title New Dramaturgies gestures towards our interest in discussing contemporary and future practices, yet the series is also concerned with new approaches to performance histories, always considering these in vibrant relationship to what is happening in the present, both in terms of artistic and wider cultural developments.

Exeter, UK Cathy Turner
Stockholm, Sweden Synne Behrndt

Acknowledgements

I would like to acknowledge the lands, traditional territories and Nations in which this writing, practice and research has taken place. This book was written in Reading, UK and involves research from multiple locations and many traditional territories (noted in the text). I acknowledge that universities and institutions in the Global North, including the one which I work for in the UK, were built on colonialism, slavery and empire, the exploitation of people and land. These contexts are a shared history and ongoing reality. I acknowledge the violent practices in which land and resources were stolen from Indigenous peoples and other First Peoples, as well as the languages and knowledges lost through the imposition of Euro-Western frames. Settler colonialism and neocolonialism continues to violently reverberate across lands, oceans and peoples today.

Thanks go to many people for their contribution to the process of creating this book. First and foremost, I would like to thank Sally Mackey, a generous, thoughtful, frank and kind PhD supervisor and mentor, who taught me how to be an academic. I am also grateful to Gareth White for his feedback and support at the very beginning when this project was in PhD form.

I am so appreciative of Adelina Ong for her honest and thoughtful feedback and for being an inspiring research companion, for listening and prompting me always to think further and resist authority. Thanks for all the days spent at the British Library. My thanks also go to the book support group: Angie Spalink and Jonah Winn-Lenetsky, whose comments and friendship were invaluable to developing this book. I am also grateful to Tanja Beer, Courtney Ryan and Bronwyn Preece, who gave me generous feedback. Special thanks go to Tamara Courage, John Whitney and Sarah Wray for their proof-reading, and to Tanya Izzard for indexing. As for my colleagues in the Department of Film, Theatre & Television at the University of Reading, thank you for being a supportive team. Thanks especially to Anna McMullan and Vicky Angelaki for their guidance at the proposal stage; Lucy Tyler and Adam O'Brien for their thoughtful comments on chapters and continued support; Jonathan Bignell, Lisa Purse and Faye Woods for helping to carve out time to complete this project. Financial research assistance was gratefully received from the Heritage and Creativity Research Theme. I'd also like to thank the third year FTT students of Performing Ecology, who engaged with these ideas as I tested them out and creatively responded in truly impressive ways.

Many thanks to my talented and inspiring collaborators Tanja Beer, Rosie Leach, Bronwyn Preece, Meghan Moe Beitiks and to all those who participated in the *Trans-Plantable Living Room*, particularly the volunteers at Riverside Community Allotments. I would also like to thank all the artists that have generously allowed me access to their work: David Harradine, Sheila Ghelani and Sue Palmer, Violeta Luna, Chantal Bilodeau, Deke Weaver and Zoe Svendsen.

I am grateful to those who have given valuable support to this research in different ways, through advice and encouragement, including Kat Low, Gilli Bush-Bailey, Sarah Standing, Sarah Blissett and Evelyn O'Malley. Theresa J. May's work inspired my early thinking on this subject and she has offered generous guidance since. Thanks also to the Ecology and Performance working group at ASTR, which has been a tremendously fruitful place for the development of this research, as well as the Theatre and Performance Research Association in the UK. Many thanks to Cathy Turner and Synne Behrndt for their insight and generous support as series editors, and to Tom Rene, Vicky Bates, Jack Heeney and Eileen Srebernik at Palgrave for guiding me through the process. I am especially grateful to my family, parents and sister for their tireless encouragement (and always asking if the book was finished yet), and especially to my partner Mike Medaglia, whose talent, love and care enabled this book.

Contents

List of Figures

CHAPTER 1

Introduction

As I write this, climate change has finally become the centre of media attention to an almost unprecedented degree.[1] In London, Extinction Rebellion activists have blocked off five busy sites, which include Oxford Circus and Waterloo Bridge, over a week in an act of non-violent civil disobedience in April 2019. They have three simple demands: they want the government to 'tell the truth' about the climate emergency; they want emissions cut to net zero by 2025; and they want a citizen assembly to inform climate-related decisions (Extinction Rebellion n.d.). That same week, David Attenborough's documentary *Climate Change—The Facts* (2019) has aired on the BBC and internationally. The destructive and fatal reality of climate change is unequivocally presented by one of the most authoritative voices on nature in the UK. During this time, young climate activist and founder of the school strike movement Greta Thunberg has delivered a speech to European Union leaders holding back tears: 'I want you to act as if the house was on fire…We are in the midst of the sixth mass extinction' (quoted in Rankin 2019). Here, climate change has finally been placed front and centre in the media, having pushed Brexit out of the UK headlines for at least a week. Perhaps a critical mass has been reached. There is no longer debate on whether anthropogenic climate change is happening and at what rate, rather what the effects will be, emissions reduction targets and what other action is necessary. The whiteness of these voices have joined Inuit voices, African voices, Central American voices, countries like the Maldives, Bangladesh and the Marshall Islands,

L. Woynarski, *Ecodramaturgies*, New Dramaturgies,
https://doi.org/10.1007/978-3-030-55853-6_1

and other Indigenous[2] voices who have all been saying for years that urgent action is required as they are already experiencing the life-altering and at times deadly effects of global climate change. This is the ecological (and social, political) context in which this book has been written.

The urgency of climate change and its associated effects have created an ecological imperative for all fields to address. As theatre scholar Wendy Arons asserts: 'humanity's relationship to the environment is an issue of urgent concern, and one that can and should be addressed by anyone engaged in critical and intellectual pursuits, including theatre artists and scholars' (2007: 93). Theatre and performance can offer something *distinctive* in their engagement with ecology. They can upend reductive narratives and images, embodying and performing contradictions, erasures and imaginative possibilities. Like theatre scholar Carl Lavery (2018), I am skeptical of hyperbolic claims of what theatre can do, particularly in relation to behaviour change. The problem–solution model, drawn on when theatre is utilised to 'communicate' specific ecological problems and 'solutions', often instrumentalises performance in a reductive way and largely focuses on content. This approach does not always leave room for the nuance, complexity or intermeshment of contemporary ecological issues. Rather, my argument that theatre and performance can offer new frames of thinking, feeling and viewing, or tell/show us something about our current ecological situation, follows theorisations on the social impact of theatre and performance in relation to ecology from such thinkers as Heddon and Mackey (2012), Arons and May (2012, 2013), Kershaw (2007) and Allen and Preece (2015). As Deirdre Heddon and Sally Mackey suggest, 'it is the combination of artistry and reality, of aesthetics and world, that has the potential to produce affect' (2012: 176). Through these combinations, theatre and performance have the potential to engage ecological thinking in unique ways to other mediums, speaking to our current context. This is the animating principle of my theory of ecodramaturgies.

Our current global ecological circumstances could be characterised as a crisis, catastrophe (Morton 2010) or emergency (Emmott 2012). The United Nations Intergovernmental Panel on Climate Change has identified a tipping point of 1.5° Celsuis (above pre-industrial levels) as the level at which the impacts of climate change would be irreversible and devastating (Masson-Delmotte et al. 2018), which is likely to happen within 12 years. However, climate change is an immediate threat; it is already producing unequal and fatal effects for many. As Mike Hulme argues, 'climate change is not "a problem" waiting for "a solution". It is an environmental, cultural and political phenomenon that is reshaping the way we think about ourselves, about our societies and about humanity's place on Earth' (2010: 41). Climate change is just one aspect of ecology; however, it is the grand narrative of our current ecological context. Climate change is ecological in that it requires an urgent and radical reconsideration of the relationship between humans and the earth, how we live and how we shape, and are shaped by, the more-than-human world.[3] This calls for a close examination of how climate change intensifies inequalities and injustices, falling along familiar patterns of vulnerability and marginalisation: race, gender, class, disability, social mobility, political capital and colonisation. This book asks questions about how theatre and performance embody, reveal and intervene in these inequalities in a climate-changed world.

To understand the approach I have taken in this book, it is important that the 'I' (and my knowledge) is situated. I am a white, Canadian (with Belgian, Ukrainian, Polish and British heritage) cisgendered woman who currently resides in the United Kingdom and holds a permanent academic post at a university in an urban centre. As such the critique of dominant Western worldviews, underpinned by colonialism, in this book is also self-reflexive. My experiences have shaped my political thinking, which informs this book, as I strive for social values of justice, rooted in intersectional feminist, anti-racist, decolonial, queer, disabled and non-anthropocentric ways of thinking. This is ongoing work that I do not always get right. My aim is not to speak for any group of people, rather I want to think critically about some of the underlying ethnocentric assumptions made about ecology, with an understanding of how this might develop what Joni Adamson refers to as 'a more inclusive environmentalism and a more multicultural ecocriticism' (2001: xix) by revealing different stories and dramaturgies.

My interest in this area stems from a formative experience learning about global warming and a long-nurtured love of live theatre and performance. As an 11 year old, I remember learning about global warming being caused by greenhouse gases, and at the time in 1995, CFCs (chlorofluorocarbons) were identified as a prime culprit. I clearly remember an infographic that illustrated how greenhouse gases trapped in the atmosphere causes temperatures to rise. The hole in the ozone seemed like an immediate threat in the cultural imaginary of North America (the angel in Tony Kusher's 1991 *Angels in America* enters earth through a hole in the ozone), while global bans of ozone-depleting chemicals were passed. I was struck by the deep injustice of global warming then as it imprinted on my young mind a specific view of the future: chemicals were trapped in the atmosphere and were destroying it. 1995 was also the 25th anniversary of Earth Day and there was a certain cultural capital associated with (mostly white) environmentalism. My interest in ecological issues continued as I became involved in activist organisations and efforts during my undergraduate degree.

My environmental consciousness-raising happened separately, but in parallel to, my excitement at live theatre. I was born on traditional Anishinabewaki territory in what is now known as Cambridge, a small city in Ontario, Canada. Growing up I was privileged to be taken to see theatre regularly as a child, including community theatre and annual trips to see musicals in Toronto or plays at the Stratford Festival in Ontario. The thrill of live performance was established in me from a young age. I also had the chance to perform in summer camp productions (including delighting in a large costume closet), and then study drama in high school, where I developed my theatre-making, writing and directing my own plays and taking them to local theatre festivals. My ecological interest and my interest in theatre and performance remained separate until I wrote and directed a play about a group of young people facing an uncertain ecological future for my undergraduate honours project at the University of Guelph, finally bringing the two together. This stemmed from a rather naive belief in the social and political power of theatre and performance after studying protest and political theatre practices. Although my thinking has become more nuanced and complex since that point, after writing my MA dissertation and PhD on the subject, I am still committed to the idea that theatre and performance can speak to critical socio-political and ecological contexts and issues in imaginative ways, particularly in light of climate and environmental inequalities and injustices. In performances, I have been moved, angered, bored, scared and delighted; I have felt connected and isolated; I have rethought my perspective and discovered new things about the way the world works. It is from this position that I approach *Ecodramaturgies: Theatre, Performance and Climate Change*.

The differentiated violence of climate change is difficult to conceptualise for those who are not experiencing it directly. Rob Nixon, in his book *Slow Violence and the Environmentalism of the Poor* (2011) describes climate change and other ecological slow-moving crises as 'long dyings' (2). This is the kind of violence that is often unseen or misunderstood because it happens gradually over a long period of time, 'a violence of delayed destruction that is dispersed across time and space, an attritional violence that is typically not viewed as violence at all' (2). Nixon argues that this slow violence has consequences across geographies, race, gender and economic mobility, as 'it is those people lacking resources who are the principal casualties of slow violence' (4). Although slow moving and cumulative, it is violence nonetheless, often against marginalised peoples, species, places, and non-humans. Climate change needs to be understood as violence in order to understand its effects. Although it is slow moving, it is happening now and many people are suffering at times fatal effects of droughts, floods, fire, hurricanes, depleted soil, loss of biodiversity and polluted air. It is not a White, Western and 'full-stomach phenomenon' (Nash 1982) as people of colour in the Global South (and other places) are experiencing the violence of climate chaos on their living conditions, health, livelihoods and well-being.

Connected to the concerns that animate this book are the representation issues for conceiving and understanding the idea of the slow violence of climate change. Nixon perceives this as a representational barrier: 'how to devise arresting stories, images, and symbols adequate to the pervasive but elusive violence of delayed effects...How can we turn the long emergencies of slow violence into stories dramatic enough to rouse public sentiment and warrant political intervention' (2011: 3). These key questions are related to the central questions I am asking about images, narratives, values, themes, processes, ethics and experiences of ecodramaturgies in representing the complexity of ecological relationships across different people, places and more-than-humans. Stories of climate change are often anonymous and happen over timescales that are not compatible with our understanding of temporalities or our human-centric narrative preferences or the political cycle or the media cycle. These images, narratives and stories need to foreground the way the slow violence of ecological destruction magnifies differences, oppressions and vulnerabilities and the power structures that underpin them. A lot about the world has changed since I learned about global warming as an 11 year old, but theatre and performance have continued to find ways of engaging and interrogating the current complex reality. I hope this book contributes to the rich history of understanding the interaction between theatre and the world.

In increasingly polarised political climates (in which climate change is remarkably still a contested term in some places), I argue that theatre and performance can open up ways of seeing and thinking, reflect blind spots and injustices, nuance ecological ideas and conversations, ask questions and problematise dominant anthropocentric modes of representation.

This is based on what I call intersectional ecologies: a way of interpreting ecodramaturgical practices, foregrounding marginalised perspectives and acknowledging the multiple social and political forces that shape climate change and related ecological crises. This is the lens I use to think and write about a broad spectrum of ecologically oriented performance practices, largely from the UK, US, Canada, Europe, Africa and Mexico from predominantly 2007-onwards. The focus on the contemporary allows for exploration of timely ecological questions, including persistent issues such as environmental justice, urbanisation, reductive images associated with whitewashing climate change and the way in which race, class, gender and colonialism produce ecological contexts and effects. Intersectional ecologies considers overlapping injustices, exclusions and oppressions asking who is affected and marginalised, and whose voice or perspective is being heard and whose is being erased. This book builds upon and expands the work of other scholars who have contributed to the burgeoning discourse of performance and ecology, extending it through the practice and critical tool of ecodramaturgy. My aim is to broaden this discourse through a variety of performance forms, including site-based, participatory, immersive, installation, activism, film, live art and text-based plays, read from an intersectional ecological perspective.

The current ecological crisis, and my own context within it, is one of the reasons I took up this research. Although I consider myself an 'ecologically conscious' person, I am often conflicted, unsure and confused about what to think, how to feel, how to move forward and how to take action. Heddon and Mackey identify uncertainty and precarity as key states when engaging in research about environmentalism and performance. They write: 'that science is so visibly unable to offer a definitive solution to climate change prompts a new and potentially productive sensibility, the acceptance of uncertainty: of epistemology, of actions, of results, of futures' (2012: 169). They link this uncertain state to the uncertain nature of performance; particularly how multiple actants of performance (such as audiences) might respond and experience a performance. Heddon and Mackey suggest that this uncertainty (about 'solutions', the future, and the best way to address the ecological crisis) may be productive and well placed in performance. Robert Butler, in his 2008 blog post on Ashdenizen, argued that one of the reasons theatre was (then) reluctant to engage with climate change was because theatre and performance-makers were unsure of what to think about it. Providing a productive edge to uncertainty, Butler wrote 'but not knowing what you think about something is the perfect moment to engage with it' (2008). It is from this position that I approached this project. Rather than eschew this complexity and uncertainty, I was motivated to take up this research because of it. The urgency of the global climate emergency will not be helped by reductive simplifications; complexity (and perhaps confusion and conflict) can be embraced as generative concepts in imagining how we live within it.

I am inspired by Linda Tuhiwai Smith's thinking on decolonising methodologies, mindful of how 'research' and 'scholarship' can be used (and have historically been used) to further the ideological supremacy of the 'West', often setting up the Other as anyone not a White, Western man. After Edward Said, Tuhiwai Smith argues that the idea of the Other is constructed through Western scholarship, institutions, vocabulary and ideological discourses often underpinned by colonialism. 'Both the formal scholarly pursuits of knowledge and the informal, imaginative, anecdotal construction of the Other are intertwined with each other and with the activity of research' (Tuhiwai Smith 1999: 2). In this book, I aim to avoid the colonising legacy of research which also constructs the more-than-human as Other, in addition to women, people of colour, Indigenous peoples, people with disabilities and many more. I position research as Tuhiwai Smith does, 'as a significant site of struggle between the interests and ways of knowing of the West and the interests and ways of resisting the Other' (Tuhiwai Smith 1999: 2), extending the Other to include more-than-human actants. This book aims to highlight forgotten, erased or marginalised experiences and narratives by some of those most affected by ecological crisis.

Conceptual Background

Before discussing where this project sits within the conceptual field of performance and ecology, it is useful to clarify some of the terms used. 'Ecology' describes the interconnected relationships of the living world, 'the study of animals and plants, our habitat and environment, as well as the analysis of the interrelationships between us all' (Giannachi and Stewart 2005: 20). The term refers not to the biological scientific study of organisms in the environment but the way in which we as human beings relate to each other, our environment, and the more-than-human world. Di Battista et al. indicate the complexity: 'Ecology speaks to both complex webs of relations between the human and non-human—themselves ideological, racialized and problematic conceptual markers—and the simultaneously fraught and comforting notion of "home", the *oikos*' (Di Battista et al. 2015: 3). I discuss home and *oikos* in detail in Chap. 5. Ecology, in this sense, focuses on the experience of ecological relationships in our everyday lives, in lieu of ecological science. Allen and Preece reflect on the limitations of term: '"Ecology" is of course a labile term, and an inherently western notion: in its very usage there is an entrenched exclusivity, which favours certain world-views over others. It is also a word that is currently enjoying much popular (mis)use in an array of different contexts' (Allen and Preece 2015: 5). As Allen and Preece articulate, the idea of ecology is contingent and situated with Western culture, to the detriment perhaps of other cultural ideas and worldviews.

Taking ecology out of the field of biological sciences and bringing it into a discourse of art and performance implies a different kind of knowledge and engagement. A creative and imaginative involvement with ecology can open it up, creating different modes of engagement, which in turn could give rise to new ways of thinking and making performance ecologically. For Morton, ecological thought 'isn't just to do with the sciences of ecology. Ecological thinking is to do with art, philosophy, literature, music, and culture' as 'ecology includes all the ways we imagine how we live together. Ecology is profoundly about coexistence' (2010: 4). Art, philosophy and performance can reveal, question and imagine 'how we live together' in differentiated and unequal ways. Although all art can be considered ecological in its material form, 'ecological art, and the ecological-ness of all art, isn't just *about* something (trees, mountains, animals, pollution, and so forth). Ecological art *is* something, or maybe it *does* something' (11). The theatre and performance practices discussed here 'do something': reveal, disclose, critique, problematise and extend thinking of ecological relationships in one way or another. I have approached the intersection of performance and ecology from this position of ecological thinking, which includes thinking about everyday relationships with the more-than-human, and how they are shaped by global capitalism, colonialism, ideology, race, class, gender, access and environmental injustice. Rather than collapse differences in an easy mantra about connectedness, my version of ecological thinking focuses on the ways in which these connections, and the effects of these connections, are violently unequal and disproportionate. Ecological thinking is implicit throughout my analysis, as I theorise the way in which ecodramaturgies are interconnected to the material world, and the way in which these asymmetrical interconnections are opened up in and through performance.

Like ecology, the dominant understanding of dramaturgy in 'Western theatre' is entrenched in a Eurocentric frame, as Dione Joseph suggests: 'within a Eurocentric paradigm, understandings of dramaturgy can be traced to secularization of the form in the sixteenth century and reflect the critical examination of a play's structure and organization as well as input into the development of the process of the play' (Joseph 2019: 131). In response to this representational paradigm, a decolonising dramaturgy, for Ric Knowles, is one 'in which performances are structured through and grounded in embodied understandings of worldviews that are deeply encoded in culturally specific cultural texts, performances, and practices' (2015: 37). As Cathy Turner points out, dramaturgy has now expanded into diverse directions, speaking to forms beyond the single-authored text-based theatre which have opened it up:

> Post-dramatic theatre and live art practices encourage us to broaden our understanding of dramaturgy beyond conceptions of the "drama" and synthesis of meaning, to encompass processual and open-ended structures, admitting the aleatory, entropic and chaotic, examining the potential for multiple narratives, frames and forms of textuality. (Turner 2010: 150)

I think about dramaturgy as a holistic approach to the way theatre and performance make meaning, which can be through analysis of play texts, but also through other elements of performance as Turner suggests: 'dramaturgy is as likely to be concerned with the use of space, visual elements, sound, audience proxemics and other aspects that might be less directly addressed by play texts' (Turner 2015: 3). My focus is on how all of these meaning-making elements relate to ecology; the integration of some of the more formal qualities with the social context of the work. Dramaturgy is expanded in this way to consider how ecological thinking is enacted, embodied and performed through ways of viewing, making and experiencing performances.

Theatre scholar Theresa J. May coined the term ecodramaturgy, describing it as 'theatre and performance making that puts ecological reciprocity and community at the center of its theatrical and thematic intent' (in Arons and May 2012: 4). May's foundational work applies the term predominantly to play texts and intent, attending to the injustices of ecological crisis. I extend it to an analysis of performance more broadly, thinking about meaning-making strategies, in a variety of performance forms, in relation to ecology. My theorisation of ecodramaturgies considers performance forms, themes, processes, narratives, values, politics, ethics and experiences. For Eckersall, Monaghan and Beddie, 'to think of dramaturgy in terms of ecology foregrounds the crucial importance of connectivity, of relationships between people, objects, natural forces and their interaction in the human/natural environment' (2014: 20). I argue that ecodramaturgies can subvert dominant forms of representation that often reduce and devalue the more-than-human world and ecological effects on people. This echoes Arden Thomas' thinking: 'by stirring the collective imagination towards a deeper sense of our material embeddedness in and accountability for the ecomaterial world, ecodramaturgical practices are poised to shift the paradigms of human-nature relations and to change audience perceptions of themselves. With its insistent emphasis on embodied connectivity, performance practices are crucial sites of investigation into the networks of exchange between culture, the environment and animals' (Thomas 2016: 201). Ecodramaturgies are often activist or gestures of resistance through diverse practices, forms and responses. They are inherently political in their responsiveness to socio-political contexts.

Ecodramaturgies are a way of understanding how theatre and performance practices make ecological meaning and interact with the material more-than-human world, attendant to the different experiences, complexities and injustices that entails. This book considers how different dramaturgies are concerned with ecology, putting forward various ways of thinking about ecodramaturgies. Thinking ecologically requires a shift in perspective to decentre the human, question neoliberal environmental logic and reimagine the nature/culture binary. In theorising the potential and possibilities of ecodramaturgies, this book aims to address the following questions: What dramaturgical strategies offer new ways of thinking about material encounters with the world, critique anthropocentric neoliberal binaries or make meaningful marginalised forms of ecological knowledge and worldviews? How can theatre and performance reveal the way different bodies (human and non-human) are exposed through environmental injustice and unequal climate change effects? How can theatre and performance, potentially through erasure or omission of places or people, throw into relief the multiple and intersecting forms of oppression and marginalisation connected to ecology? How can an intersectional analysis open up new ways of thinking about ecological performance? I think through these questions in subsequent chapters in order to establish a set of examples for the potentialities of ecodramaturgies.

This theorisation and articulation of ecodramaturgies resists totalisation. Rather, it focuses on complexity, entanglement, tensions, contradictions and uneasily reconcilable ways of being and thinking. They are not intended to be descriptive; they are simply an attempt to advocate for diverse strategies across multiple forms. An ecodramaturgical analysis considers modes of viewing and making, narratives, values, politics, ethics and affect in process, production and reception.

The theatre and performance works included in this book are performance events, practices and plays that engage with ecology, *thematically, experientially* and/or *performatively*. In the broadest terms, these works prompt the audience (or participants) to consider ecological relationships through the content, form and/or the experience of the performance, as well as the way in which the performance enacts ecological thinking. This encompasses different forms including dance, live art, music, installation art, film, theatrical performance, eco-activism and text-based performance. In my development of ecodramaturgies, I am interested in questions of site, text, spectatorship, representation, cultural context, form, participation, scenography and space (Bottoms 2003) from an intersectional ecological point of view.

Semantically, I do not make a distinction between theatre and performance per se, opting for an expanded idea of theatre. Instead of being associated with the 'pre-scripted blackbox', or 'linguistic artefact', I note Lavery's emphasis on the performative:

> in contemporary practice and theory, theatre is seen as a predominantly performative medium, that is to say, as something embodied, ephemeral and affective, with the result that the fundamental concern of scholars is no longer to decipher what the theatre text means but rather to focus on what the theatre medium 'does'; in how, that is, its dramaturgical distribution of organic and inorganic bodies in actual time and space creates sensations and experiences in the here and now. (2016: 230)

I theorise all of the works in this book as 'doing something' in relation to ecological thinking, including revealing ecological relationships, critiquing specific practices or our relationship to the more-than-human world, and/or deconstructing binaries between human/non-human. These works extend, problematise and/or offer a new way of thinking to my theorisation. They are often urban, resisting the anti-urban bias in much ecological work (Harvey 1993) and challenging the commonly held perception of ecology as 'green and pleasant'. The body of work I include also addresses a gap in scholarship around the spectrum and diversity of ecologically oriented theatre and performance work. However, they are not an exhaustive and complete index of all ecological theatre and performance. The broad spectrum of work is suggestive of the possibilities of ecodramaturgies, indicative of the diverse forms of practices that this term might encompass.

Ecological Waves

Broadly, performance and ecology bring ecological thinking to bear on theatre and performance criticism, dramaturgy, production and scholarship. The field, or emergent area of study, as asserted by eco-performance philosopher Wallace Heim, is multivocal and not categorically recognisable, rather 'a family with resemblances, always changing, with its borders hazy and continually porous to other kinds of art-making and other sources of knowledge and experience' (Heim 2014: 6). I think through the historical background of the field, loosely connecting the 'family resemblances' into three waves. It is important to note the 'waves' of scholarship in performance and ecology detailed here are mainly Anglocentric, English-speaking Western, emanating primarily from the North American, UK, Australian and New Zealand academic contexts, which means it is a restricted and limited view of ecology and the broad range of theatre and performances practices that engage with it. The wave analogy is useful but imperfect as it does not necessarily take into account that theoretical strands develop in complex and tangled ways and do not fit neatly into a bounded timeframe. Yet, waves are helpful as they leave behind traces, which might get picked up by a further wave. The constant movement and circulation allow for ideas to merge and interact in generative ways. Not all work in this area can be neatly categorised into three distinct phases; rather, ideas are circulating, with the boundaries between the waves being usefully liminal. Noting the limitations, the three 'waves' of this work is a helpful analogy for the historisation and conceptual background of this book.

Performance and ecology as a topic of study has been developing in earnest since the mid-1990s. Since that time a lot about our ecological situation has changed. In 1994, global warming was seen as a relatively serious threat, as the United Nations Framework Convention on Climate Change was agreed in Rio de Janeiro; however, it was still considered distant and slow moving so drastic action was not widely taken. The first wave of performance and ecology work initiated around this time with a special issue of the journal *Theater* is characterised by articulating the need of such a discourse, as well as evoking ecocriticism tropes, particularly romanticism and wilderness ideals and the metaphorical use of ecology. The first wave also provided some early theorisations of the role community-based performance might play in engaging with ecological concepts and concerns.

This first wave of scholarship is prefigured by political and social theatre practices (with North American and UK groups having well-documented histories). Activist groups, political theatre companies and performance artists employed eco-performance practices in response to ecological concerns and the growing environmental movement of the 1960s and early 1970s. Activist groups such as Greenpeace and Earth First! used performance strategies to gain media attention and further their causes. The emblematic Bread and Puppet, a US-based political theatre company known for creating circus-style performances with large-scale puppets in a farm in Vermont, also engaged with ecological issues. For example, their mid-1990s work opposing the removal of community gardens in NYC for 'development' under then-Mayor Giuliani, included puppets of the gardens and the often-used Mother Earth character (Bell 1998: 272). In the UK, there were counter-culture radical theatre groups also making work about ecological issues of the day, such as the Campaign for Nuclear Disarmament, oil drilling, and displacement and land rights. Many of these alternative companies were community-orientated and aimed to make work that would affect local change (Kershaw 1992: 5). Other theatre and performance companies in the US and South Pacific in the 1980s and 1990s engaged with ecological activism, as community-based performance or children's performance including Precipice Theatre, The Wan Smolbag Theatre, Evergreen Theatre Society and Theatre in the Wild. The political theatre roots of ecological performance used strategies such as reframing sites, resistance, spectacle, audience participation and community engagement to enact ecological thinking in different ways. These strategies are utilised by some of the contemporary theatre and performance works that I focus on in this book.

John McGrath's *The Cheviot, the Stag and the Black, Black Oil* (1973) is an example of the way in which radical performance engaged with ecological issues of the time. Made by 7:84 (Scotland) performance group, the play tells the continuing story of Highland history marked by exploitation: first the nineteenth century forced migration of rural Highlanders to claim the land for Cheviot sheep farming by aristocratic landowners, then the wealthy estate owners using the land for stag-hunting and finally the development of the North Sea oil fields. The performance toured 27 community venues in the Highlands of Scotland in 1973. The show was a ceilidh play, full of songs, storytelling, jokes and dancing and 'aimed to illuminate the situation, and to recommend radical action for the future' (Kershaw 1992: 152). The play tied together capitalist exploitation, class oppression and ecological issues. The popularity of the play, which told an alternative history and questioned the future of Scotland, was emblematic of the kind of community-orientated and socio-political-ecological performance that was being made at the time. These early activist performances built an activist basis for ecological performance in the Anglo-west that continues today. The lineage of these ecological theatre and performance works that have been able to galvanise communities around ecological issues, and were attendant to some of the intersectional concerns of class, gender, neoliberal capitalism and urbanisation still ongoing today.

Early thinking in performance and ecology scholarship developed from ecocriticism in literature. Cheryll Glotfelty describes ecocriticism as 'the study of the relationship between literature and the physical environment ... ecocriticism takes an earth-centered approach to literary studies' (1996: xviii). Performance and ecology could be thought of as taking an earth-centred approach to theatre and performance, viewing performance as part of the larger world which 'does not float above the material world in some aesthetic ether, but, rather, plays a part in an immensely complex global system, in which energy, matter, *and ideas* interact' (xix). This thinking is useful in relation to performance and ecology, and in perhaps reconfiguring the binary that relegates theatre and performance to 'culture' (as opposed to nature). Theatre and performance do not exist in a vacuum; they impact and are impacted by the world of social-ecological systems and relationships. Ecodramaturgies speak to this relationship by asking how theatre interacts with the world.

Una Chaudhuri's landmark article, '"There Must Be a Lot of Fish in That Lake": Toward an Ecological Theater' (1994), (in a special issue of *Theater* also featuring articles by Munk, Bell, Fuchs, Barnett and Rabillard) called for a move towards an ecological theatre.[4] Chaudhuri issues an important call to arms, contending that theatre must engage with ecology for its 'own standards of social seriousness and political relevance' and because 'ecological victory will require a transvaluation so profound as to be nearly unimaginable at present. And in this the arts and humanities—including the theatre—must play a role' (25). She identifies the ecological crisis as a crisis of values and the theatre 'as the site of both ecological alienation and potential ecological consciousness' (25). Although I conceptualise ecological thinking differently, as intersectional, relational, entangled, messy, contingent and in flux, this work set up a useful foundation for thinking about how theatre and performance can be thought of through ecological frameworks.

Writing before the new millennium, Chaudhuri further contends that the Western theatrical imagination of the last century is 'haunted' by 'the ticking time bombs of ecological disaster' (1994: 23). At this time the most common forms of ecological theatre are ones that include 'an underlying and dystopic ecological condition pervading the world of the play' (23), which have largely remained unexamined (i.e. Beckett's *Endgame* (1957), Ibsen's *An Enemy of the People* (1882), Rivera's *Marisol* (1992), which all enact a negative ecological vision). One of the reasons for this, she argues, is the humanist foundation on which modern drama, namely naturalism and realism, is built. Naturalism and realism put forward a 'wholly social account of human life' (24), seemingly divorcing humans and culture from the natural world. Chaudhuri even goes as far to say that naturalism is 'anti-nature', despite the fact that it is teeming with ecological imagery:

> Though its thematics kept in touch with nature through images of cherry orchards, wild ducks, and polluted baths, the ideological discourse of realism thrust the nonhuman world into the shadows ...The junk-strewn, garbage-choked stages of Pinter, Mamet, Shepard, and others, reveal naturalism's anxiety—long concealed—about the widening gap between the human and the nonhuman. (1994: 24)

The first step towards an ecological theatre, according to Chaudhuri, is to acknowledge the rupture between humans and nature that theatre participates in.[5] By making space for this acknowledgement, 'the theatre can become the site of a much-needed ecological consciousness' (28). Many of the performances I write about in this book reframe or subvert this binary between humans/nature and nature/culture. It is these dualisms which have ideologically shaped the relationship to the ecological world in the Global North that performance might reimagine in much more complex ways.

The first wave of performance and ecology also included Downing Cless (1996) who disagreed with Chaudhuri's assertion, contending that the humanist tradition of the theatre does not necessarily have to be in opposition to ecology. Cless (2010, 2012) advocates the role of the director in staging more ecological interpretations of works (such as Shakespeare's *A Midsummer Night's Dream*, Beckett's *Waiting for Godot* and Chekov's *The Cherry Orchard*) or offering an ecocritical reading of them. Also among some of the first-wave writings was Bonnie Marranca's *Ecologies of Theater* (1996). Although Marranca sets out an aim to outline a biocentric worldview, she employs ecology broadly as a metaphor for theorising avant-garde, experimental performances. In my thinking, ecology as a metaphor or analogy for any type of system or network tends to bracket off the relationship to political and material reality. Given the urgency of ecological issues, within this book I use the term ecology in the material rather than metaphorical sense.

Erika Munk's 1994, or 'first wave', plea for theatre and performance to develop rich and rigorous histories and theories of the field 'went largely unheeded until recently, when a smattering of articles and books appeared on the subject ... which have helped lay the groundwork for a recognizable, growing discourse, grounded in a variety of methodologies, from theatre historiography to performance studies' (Arons and May 2012: 4). This 'smattering of article and books', roughly between 1999 and 2010, is what I characterise as the second wave of performance and ecology scholarship. The landmark conference 'Between Nature: Explorations on the Edge of Ecology and Performance', held at Lancaster University in the UK in 2000, brought together 200 international participants to explore the connections between culture, ecology and performance. It spawned collaborations and joint projects, including two edited collections:[6] *Nature Performed* (Szerszynski et al. 2004) and *Performing Nature* (Giannachi and Stewart 2005), which is one of the first English-language edited collections to specifically explore ecology and the performing arts. The formation of the Earth Matters on Stage (EMOS) organisation and symposium at Humboldt State University in California in 2004 created a further wealth of thinking on the subject and more artistic opportunities with an eco-drama playwriting competition. Founded by Theresa J. May and Larry Fried, EMOS is 'a consortium of artists, educators, activists and scholars who believe that theatre and the performing arts must respond to the environmental crisis...The EMOS symposium fosters dialogue about the intersection of environment, culture, and performance' (Earth Matters on Stage).[7] Heim's establishment of the Ashden Directory, a database of ecologically engaged productions with news, a timeline, interviews and the Ashdenizen blog, active from 2000 to 2014, was about 'endorsing the possibilities for performance to change people's perceptions of what it means to be human and interdependent with environments and with nature, with other-than-human' (Heim 2014: 6). The formation of these three organisations and resources, laid the foundation for the more diverse and multivocal work that would come out of the third wave.

In her influential article, 'Greening the Theater: Taking Ecocriticism from Page to Stage' (2005), May highlights the need to have an intersectional approach to environmental issues (although 'intersectionality' as a term was not yet in popular usage), calling for ecocriticism to be pushed beyond the white, male wilderness aesthetic to consider how environmental degradation disproportionately affects the poor and communities of colour. 'When the ecocritical view can expand its scope to include the issues of race, class, gender, geographic situated-ness, and white power and privilege, then theatre—which has always been a force for activism as well as the dissemination of hegemonic myths—appears ripe for analysis' (2005: 87). She goes on to analyse the way American theatre perpetuated the narrative of frontierism at the turn of the century and participated in the subjugation of Indigenous peoples, while also objectifying the landscape. It is from this spirit that I base my analysis, recognising that ecological thinking means unpacking issues of privilege, white supremacy, class, gender, sexuality and Indigeneity. These animating ideas run through this book, although I discuss them in further detail in Chap. 2, with intersectional ecologies, and Chap. 6, where Indigenous ecodramaturgies are the focus.

Despite the works and events mentioned above, the second wave was still largely characterised by a relative absence of scholarship and research that took performance and ecology seriously, perhaps as ideas were gestating. This echoes Wendy Arons' argument that theatre and performance scholars, for the most part, have been slow to engage with ecology (2007). This lack of engagement may be due to the culturally inherited separation between nature and culture 'that keeps theatre scholars trapped in binary ways of thinking about what performance is and does' (93). She further suggests it could be due to the field of study seeming somehow uncritical or less sophisticated than other topics of interest to theatre and performance studies: 'the presumption that the intersection of ecological concerns and theatre produces the kind of amateur, sincere, heartfelt, preaching-to-the-converted production' (93). These are not necessarily the kinds of theatre and performance practices I discuss in this book, although I think they are worthy of study in their own right and want to resist any high/low art binaries that this implies. Ecology and climate change are no longer 'fringe' issues, but ones that are already affecting the everyday lives of billions of people, animals and plant species. These concerns are becoming increasingly urgent and apply to everyone, in every discipline, and should be engaged with through a range of different modes and practices.

The third wave of performance and ecology is characterised by a more rhizomatic structure, with diffuse and diverse areas of study developing simultaneously, inflected and inspired by contemporary social and political issues. Between 2010 and the beginning of 2020, there have been no less than 25 edited collections, books and special issues of journals in English on topics related to performance and ecology, demonstrating the growth of the field.[8] These works set out a renewed critical agenda for the third wave. As Arons and May contend: 'what was a fringe conversation in theatre studies only 15 years ago has burgeoned into a complex, nuanced, diverse, and multivoiced praxis' (2012: 9). Some of these directions of growth have included outdoor Shakespeare and ecomaterialism (O'Malley 2018), the ecological image (Lavery 2019), children and intergenerational performance (Hopfinger 2018), site-specific river performances with communities (Matthewman et al. 2015, Scott-Bottoms 2019), ecoscenography (Beer 2017, Hann 2020) and the ongoing effects of the BP oil spill in the Gulf of Mexico (Ryan 2019) to name only a recent few. In my characterisation of the relationship between performance and ecology, I aim for these attributes of complexity, nuance and a diverse, intersectional praxis.

This third wave has also seen a proliferation of climate change dramas[9] that have been staged in London since 2009. While there have been a number of these plays,[10] indicating the increased visibility of the ecological themes and topics in theatre, they are not the focus of my analysis. Arons and May acknowledge the difficulty in playwrights trying to dramatise stories about climate change, the scale of which is so vast and potentially beyond human forms of measurement and comprehension, that 'even when a playwright strives to foreground ecological issues on stage, the stories are hard to contain' (2012: 4). Instead of placing the onus on the playwrights, within this book I am interested in how theatre and performance enact or reveal ecological relationships, through an ecodramaturgical analysis of practice. Heddon and Mackey write about the failure of these climate change plays to engage audiences in environmental thinking 'with audiences remaining unengaged and unaffected by environmental themes. Where the balance is tilted towards the emotional narratives of human relationships, those who seek sophisticated theatre addressing climate change are disappointed' (2012: 175). They align many of these plays with Mike Hulme's (2009) 'deficit model' of communicating climate change, in which the assumption is that scientists have 'the truth', which they only need to impart to the public for them to understand climate change. Hulme argues that this assumption does not engage people in scientific discourse. Instead of a deficit model, ideas of diversity, variety, circularity and multivocality are more effective in engaging the public on climate change. Rather than a pedagogic encounter or lecture, I take up the idea that theatre and performance can open up different and diverse ways of thinking about our relationship to ecology, through artistic and creative modes of engagement. It is theatre and performance that can embody and enact ecological thinking.

Despite the fact that the field is growing within theatre and performance studies, ecocriticism and environmental humanities, there is still a need for further scholarship and arts practice engaged in ecological thinking to be expanded to include more diversity (particularly ethnic, racial and geographic in terms of scholarship), in order to understand the potentialities of performance in engaging with this timely concern. By detailing performances that represent marginalised humans and more-than-humans, this book aims to contribute and expand the discourse further, building on the foundational work that has already been done.

Chapter Structure

In Chap. 2, I set out the theoretical foundations of my thinking on eco-dramaturgies as intersectional ecologies. This approach is inspired by the Green Belt tree planting movement in Kenya, led by Wangari Maathai, which built a broad base of support and influenced Kenyan politics by connecting environmental issues to gender rights, political freedoms and colonial land management. Intersectional ecological thinking is both a way of looking and praxis. It is based on the idea that, on a global scale, ecological effects are unevenly disrupted and tied to social structures that disproportionately affect marginalised people such as women, people of colour, Indigenous peoples and the poor. Forgotten histories and the ongoing ecological effects of British colonialism in India are uncovered in the story of objects in the show-and-tell performance *Common Salt*

(2018–2020) by Sheila Ghelani and Sue Palmer. Salt also functions as an ecological material and metaphor in *Salt.* (2016–2019) by Selina Thompson as she retraces the transatlantic slave trade, revealing the hidden connections between colonialism, slavery, bodies and ecological elements. *Then She Said It* (2002) by Osonye Tess Onwueme draws out the intersectional effects of a different global commodity as it focuses on oil extraction in the Niger Delta and the rural women carrying the burden of the effects. Through these works, I argue that ecodramaturgies can bring to light ecological injustices in theatre and performance through an approach of intersectional ecologies. This line of thinking is carried through in my theorisation of ecodramaturgies in subsequent chapters.

In Chap. 3, I argue for *bioperformativity* as an ecodramaturgical strategy for performance involving the non-human—one that queers ecology, disrupting the anthropocentric hierarchy through recognition of the capacity for agency and action in the more-than-human. My concept of bioperformativity is grounded in ecomaterialism and refers to the performativity of more-than-human as lively material. Bioperformativity is enacted in the ecological agency of trees, particularly in Joseph Beuys' *7000 Oaks* (1982). UK performance company Fevered Sleep's *It's the Skin You're Living In* (2013) enacts bioperformativity and queer ecological thinking through the way it troubles distinctions between human, animal and climate in its structure, narrative, editing and images. Lucy + Jorge Orta's *Symphony for Absent Wildlife* (2014) and Fevered Sleep's *The Weather Factory* (2010) are analysed as affective metaphors for the way in which a non-anthropocentric sensibility can be communicated through performance practice particularly through their staging and scenography. These theatre and performance works challenge dominant anthropocentric representational modes and values through metaphors, images, forms, processes and material effects, adding further texture to ecodramaturgies.

Following that, in Chap. 4, I argue that theatre and performance can engage the body as in and of ecological matter and material, drawing out new frames of thinking about bodily relationships and encounters. Ecodramaturgies are theorised as an embodied practice that resonates and situates the performer/audience/participant within complex 'networks of risk, harm, culpability and responsibility' (Alaimo 2016: 3) as well as desire and pleasure. The ecological body is analysed in Mexican performance artist Violeta Luna's work *NK603: Action for Performer & e-Maiz* (2014) and Cherríe Moraga's *Heroes and Saints* (1994) to consider how environmental injustices are felt in the body. *Still Dance with Anna Halprin*

(Stubblefield 1997–2000) and Ana Mendieta's *Siluetas Series* (1973–1980), further reveal the tension between bodies as 'natural', entangled, and inscribed with neoliberal values, gendered, raced and aged. Deke Weaver's immersive performance *Wolf* (2013) works to collapse separations between human and 'nature' and human and 'animal', playing with familiar stories and tropes to draw out its ecological potency. Participation in global environmental politics in Rimini Protokoll's *World Climate Change Conference* (2014) revisions some of the power structures that shape responses to climate change.

In Chap. 5, I think about how ecodramaturgies engage ideas of the Earth as home, which can be a hypothetical, imaginative, literal or material place. They can complicate binaries of local/global, public/private, home/outside world, drawing out racialised and gendered ideas of homes. Fevered Sleep's *Above Me the Wide Blue Sky* (2013) and my own practice, *Trans-Plantable Living Room* (2013) both grapple with ideas of place, home and ecology revealing the dangers and complexities of nostalgia in the context of climate change. A recontextualisation of text, characters and place setting in *A Raisin in the Sun* (1959) by Lorraine Hansberry draws on bell hooks' idea of 'homeplace making' (1990) as an entanglement of complex ecological relations of home that are also racialised, gendered and shaped by socio-political forces. As a counterpoint to home as a local idea of place, I suggest ecology is inherently bound up with home in a global context. Ecodramaturgies offer the potential to interrogate the relationship between the local and global, and heterogeneous ecological relationships and networks revealed in performance, specifically in form, structure and images in Sharon Switzer's *#crazyweather* (2013) and Shonni Enelow's *Carla and Lewis* (2014), extending the concept of eco-cosmopolitanism (Heise 2008).

Chapter 6, Decolonised Ecologies, further nuances ecodramaturgies through a critique of the Anthropocene and the way in which it erases difference and the unequal effects of climate change. I argue that theatre and performance works can foreground and reflect this critique, particularly through Indigenous ecodramaturgies. Chap. 6 attends to the specific intersections of colonialism and the Anthropocene discourse and the way Indigenous ecodramaturgies are a way of decolonising the dominant idea of the human as responsible for climate change. Colonialism has been suggested as one of the key markers to pinpoint the shift in epochs to the Anthropocene, placing the exploitation of Indigenous peoples and lands at the heart of the concept (Lewis and Maslin 2015). Indigenous ecodramaturgies are considered as a way of critiquing Western binaries between human and non-human as well as revealing the way settler colonialist domination has had shared material effects on both environments and Indigenous communities. As a non-Indigenous person, I centre the voices and writing of Indigenous thinkers, activists and artists. A mix of Indigenous activist movements, including No Dakota Access Pipeline (NODAPL) and Idle No More, are analysed alongside the plays *Burning Vision* (Clements 2003), *Salmon is Everything* (May and Klamath Theatre Project 2014) and *Sila* (Bilodeau 2015) to foreground marginalised and erased ecological worldviews and traditional knowledge and different possibilities for what it means to be human.

To conclude, I discuss the ways in which ecodramaturgies can open up a space for imagining other ways of living together, exploring alternatives to capitalist logic of extractivism.[11] The dialogue-based performance, *WE KNOW NOT WHAT WE MAY BE* (Metis 2018), created a space to imagine a future based on principles of equality, ecological justice, citizen engagement and collective decision making. Audiences were able to talk to each other and make decisions about how they wanted the future to play out and then experience what the future might be like. A future based on low-carbon living, circularity and regeneration rather than consumerism and wealth accumulation was envisioned through audience participation. Ecodramaturgies can perform other ways of looking at the future, imagining what it might mean to not accept the story of inevitable extinction, but playing with and performing potentialities and possibilities of living better together.

Theatre and performance are a part of wider cultural responses to rapidly changing ecological realities. Artistic and cultural practices and interventions are crucial to both understanding the complexities and injustices of our current context and imagining alternatives or what resistance looks like. I hope this book will make a contribution to these essential works.

Notes

1. The final stage of editing this book happened while the UK was in lockdown during the COVID-19 global pandemic in Spring 2020. The media, news and circumstances have changed drastically from a year ago.
2. Indigenous is capitalised throughout to refer to the identity 'Indigenous peoples'.
3. David Abram's (1997) term.

4. Chaudhuri's 1994 article was predated by Lynn Jacobson's 'Green Theatre: Confessions of an Eco-Reporter' in *American Theater* in 1992. *Greening up Our Houses: a guide to a more ecologically sound theatre* (1994) by Fried and May was also part of the first wave.
5. After the first wave, Chaudhuri has continued to be a leading voice in Performance and ecology. See Chaudhuri (1995, 2002, 2003, 2007, 2012, 2013, 2014, 2015), Chaudhuri and Enelow (2006, 2014), Fuchs and Chaudhuri (2002), Chaudhuri and Hughes (2014)
6. As the second wave built momentum, special editions of journals on the subject were published including the *Journal of Dramatic Theory and Criticism* (Spring 2006) and *Theatre Topics* (Fall 2007); articles including Standing (2005), May (1999, 2005), Kershaw (2002) as well as the books *Small Acts of Repair: Performance, Ecology and Goat Island* (Bottoms and Goulish 2007), *Theatre Ecology: Environments and Performance Events* (Kershaw 2007), *Green Shakespeare: From Ecopolitics to Ecocriticism* (Egan 2006) and *Green Theatre: Promoting Ecological Preservation and Advancing the Sustainability of Humanity and Nature* (Heinlein 2007).
7. Each gathering is anchored by an Ecodrama festival and playwrights competition, which is an open competition to find new eco plays, with the winner and first runner-up receiving workshop productions or staged readings. My work departs from this emphasis on playwrights to develop new work on the ecological themes and topics as I am interested in the experience and dramaturgical relationships in performance. The 2004 festival and symposium were followed up with a 10-day event at the University of Oregon in 2009 with keynotes by Una Chaudhuri and performance artist Rachel Rosenthal. The festival and symposium have continued to be held every four years in different places in the US, although to date, none have been on the scale of the 2009 event.
8. The following is a list of books and journals that were published between 2010 and early 2020: *Canadian Theatre Review* 'Theatre in an Age of Eco-crisis' (Gray and Rabillard 2010), *Ecology and Environment in European Drama* (Cless 2010), *Ecocriticism and Shakespeare: Reading Ecophobia* (Estok 2011), *Readings in Performance and Ecology* (Arons and May 2012), *Research in Drama Education* (RiDE) 'Environmentalism' (Heddon and Mackey 2012), *Performance Research* 'On Ecology' (Bottoms et al. 2012), *To Life!: Eco Art in Pursuit of a Sustainable Planet* (Weintraub 2012), *Landing Stages* (Heim and Margolies 2014); *Research Theatre, Climate Change, and the Ecocide Project* (Chaudhuri and Enelow 2014), *Enacting Nature: Ecocritical Perspectives on Indigenous Performance* (Däwes and Maufort 2014); *Performance on the Behalf of the Environment* (Besel and Blau 2014), *Rethinking the Theatre of the Absurd: Ecology, the Environment and the Greening of the Modern Stage* (Lavery and Finburgh

2015), *Ethics of Art: Ecological Turns in the Performing Arts* (Cools and Gielen 2014), *Theatre for Women's Participation in Sustainable Development* (Osnes 2013), *Playing for Time* (Neal 2015), *Performing Ethos* 'Performings Ecos' (Allen and Preece 2015), *Green Letters* 'Performance & Ecology: What can Theatre Do?' (Lavery 2016, edited collection Lavery 2018), *Building Sustainability with the Arts: Proceedings of the 2nd National EcoArts Australis Conference* (Curtis 2017), *This Contentious Storm: An Ecocritical and Performance History of King Lear* (Hamilton 2017), *Performance Research* 'On Climates' (Fensham et al. 2018), *Theatre & Environment* (Angelaki 2019), *Performance Research* 'On Mountains' (Pitches and Shearing 2019), *Canadian Theatre Review* 'Extractivism and Performance' (Richards and Davis-Fisch 2020), *Performance Research* 'On Dark Ecologies' (Spalink and Winn-Lenetsky 2020), *Theatre Research International* 'Climate Change and the Decolonized Future of Theatre' dossier (Woynarski et al. 2020), Earth Matters on Stage (May 2020), in addition to a number of national and international conferences, symposia and individual journal articles (which have been excluded here for length).

9. The Artists and Climate Change database have been keeping a list of climate change plays since 2014: https://artistsandclimatechange.com/2014/11/01/creating-a-list-of-climate-change-plays/
10. The climate change dramas on London stages since 2009 include *The Contingency Plan* (Waters, Bush Theatre 2009), *Earthquakes in London* (Bartlett, National Theatre 2010), *The Oikos Project: Protozoa* and *Oikos* (Adshead and Wu, Jellyfish Theatre 2010), *Greenland* (Buffini, Skinner, Charman, Thorne, National Theatre 2011), *The Heretic* (Bean, Royal Court 2011), *Ten Billion* (Emmott, Royal Court 2012), *2071* (Rapley and Macmillan, Royal Court 2014), *Pastoral* (Eccleshare, Soho Theatre 2013), *Lungs* (Macmillan, Paines Plough 2011, Old Vic 2019), *Mr. Burns* (Washburn, Almeida Theatre 2014), *F*ck the Polar Bears* (Ronder, Bush Theatre 2015), *Escaped Alone* (Churchill, Royal Court 2016), *Oil* (Hickson, Almeida Theatre 2016).
11. Extractivism is the ideology that the resources of the earth can and should be extracted, removed or depleted for the use of humans and for profit, particularly if the resource is in high demand (Klein 2014).

References

Abram, David. 1997. *The Spell of the Sensuous: Perception and Language in a More-Than-Human World*. 1st Vintage Books Ed. New York: Vintage Books.

Adamson, Joni. 2001. *American Indian Literature, Environmental Justice, and Ecocriticism: The Middle Place*. Tucson: University of Arizona Press.

Adshead, Kay, and Wu. Simon. 2010. *The Oikos Project: Two Plays*. London: Oberon Books Ltd.

Alaimo, Stacy. 2016. *Exposed: Environmental Politics and Pleasures in Posthuman Times*. Minneapolis: University of Minnesota Press.

Allen, Jess, and Bronwyn Preece. 2015. Decentring the Stage: Towards an Ecocentric Ethics of Performance. *Performing Ethos: International Journal of Ethics in Theatre & Performance* 5 (1): 3–15.

Angelaki, Vicky. 2019. *Theatre & Environment*. London; New York: Red Globe Press.

Arons, Wendy. 2007. Introduction to Special Section on "Performance and Ecology". *Theatre Topics* 17 (2): 93–94.

Arons, Wendy, and Theresa J. May, eds. 2012. *Readings in Performance and Ecology*. London; New York: Palgrave Macmillan.

———. 2013. Ecodramaturgy in/and Contemporary Women's Playwrighting. In *Contemporary Women Playwrights: Into the Twenty-First Century*, ed. Penny Farfan and Lesley Ferris, 181–198. London; New York: Palgrave Macmillan.

Bartlett, Mike. 2010. *Earthquakes in London*. London: Methuen Drama.

Bean, Richard. 2011. *The Heretic*. London: Oberon Books.

Beer, Tanja. 2017. Saved from the Scrapheap: Revealing the Creative and Ecological Potential of Societal Leftovers in Scenography. *Performance Research* 22 (8): 107–114.

Bell, John. 1998. Louder Than Traffic: Bread and Puppet Parades. In *Radical Street Performance: An International Anthology*, ed. Jan Cohen-Cruz, 271–281. London; New York: Routledge.

Besel, Richard D., and Jnan A. Blau, eds. 2014. *Performance on Behalf of the Environment*. Lanham: Lexington Books.

Beuys, Joseph. 1982. *7000 Oaks*. Accessed January 21, 2018. https://www.diaart.org/exhibition/exhibitions-projects/joseph-beuys-7000-oaks-site.

Bilodeau, Chantal. 2015. *Sila*. Vancouver: Talonbooks.

Bottoms, Stephen. 2003. The Efficacy/Effeminacy Braid: Unpacking the Performance Studies/Theatre Studies Dichotomy. *Theatre Topics* 13 (2): 173–187.

Bottoms, Stephen, and Matthew Goulish, eds. 2007. *Small Acts of Repair: Performance, Ecology and Goat Island*. London; New York: Routledge.

Bottoms, Stephen, Aaron Franks, and Paula Kramer. 2012. On Ecology. *Performance Research* 17 (4): 1–4.

Butler, Robert. 2008. Why Theatres Don't Touch Climate Change. *Ashdenizen* (blog), November 12. Accessed March 03, 2017. http://ashdenizen.blogspot.co.uk/2008/11/six-reasons-why-theatres-wont-touch.html.

Chaudhuri, Una. 1994. "There Must Be a Lot of Fish in That Lake": Toward an Ecological Theater. *Theater* 25 (1): 23–31.

———. 1995. *Staging Place: The Geography of Modern Drama*. Ann Arbor: University of Michigan Press.

———. 2002. Like a Tree. *Women & Performance: A Journal of Feminist Theory* 13 (1): 9–18.
———. 2003. Animal Geographies: Zooesis and the Space of Modern Drama. *Modern Drama* 46 (4): 646–662.
———. 2007. (De)Facing the Animals: Zooësis and Performance. *TDR: The Drama Review* 51 (1): 8–20.
———. 2012. The Silence of the Polar Bears: Performing (Climate) Change in the Theatre of Species. In *Readings in Performance and Ecology*, ed. Wendy Arons and Theresa J. May, 45–57. London; New York: Palgrave Macmillan.
———. 2013. Bug Bytes: Insects, Information, and Interspecies Theatricality. *Theatre Journal* 65 (3): 321–334.
———. 2014. The Sun'll Be Hotter Tomorrow: Growing Up with Climate Chaos. *Resilience: A Journal of the Environmental Humanities* 2 (1): 115–121.
———. 2015. Anthropo-Scenes: Theater and Climate Change. *Journal of Contemporary Drama in English* 3 (1): 12–27.
Chaudhuri, Una, and Shonni Enelow. 2006. Animalizing Performance, Becoming-Theatre: Inside Zooesis with The Animal Project at NYU. *Theatre Topics* 16 (1): 1–17.
———. 2014. *Research Theatre, Climate Change, and the Ecocide Project: A Casebook*. New York: Palgrave Macmillan.
Chaudhuri, Una, and Holly Hughes, eds. 2014. *Animal Acts: Performing Species Today. Critical Performances*. Ann Arbor: University of Michigan Press.
Churchill, Caryl. 2016. *Escaped Alone*. London: Nick Hern Books.
Clements, Marie. 2003. *Burning Vision*. Vancouver: Talonbooks.
Cless, Downing. 1996. Eco-Theatre, USA: The Grassroots Is Greener. *The Drama Review* 40 (2): 79–102.
———. 2010. *Ecology and Environment in European Drama*. London; New York: Routledge.
———. 2012. Ecodirecting Canonical Plays. In *Readings in Performance and Ecology*, ed. Wendy Arons and Theresa J. May, 159–168. New York: Palgrave Macmillan.
Cools, Guy, and Pascal Gielen. 2014. *The Ethics of Art: Ecological Turns in the Performing Arts*. Amsterdam: Valiz.
Curtis, David, ed. 2017. *Building Sustainability with the Arts: Proceedings of the 2nd National EcoArts Australis Conference*. Newcastle upon Tyne: Cambridge Scholars Publishing.
Däwes, Birgit, and Marc Maufort, eds. 2014. *Enacting Nature: Ecocritical Perspectives on Indigenous Performance*. Brussels: P.I.E. Peter Lang.
Di Battista, Amanda, Oded Haas, and Darren Patrick. 2015. Conversations in Queer Ecologies: An Editorial. *UnderCurrents: Journal of Critical Environmental Studies* 19: 3–5.
Earth Matters on Stage (EMOS). n.d. Earth Matters on Stage. Accessed April 30, 2019. https://www.earthmattersonstage.com.

Eccleshare, Thomas. 2013. *Pastoral.* Oberon Books Ltd.

Eckersall, Peter, Paul Monaghan, and Melanie Beddie. 2014. Dramaturgy as Ecology: A Report from The Dramaturgies Project. In *New Dramaturgy*, ed. Katalin Trencsényi and Bernadette Cochrane, 19–36. London: Bloomsbury Methuen Drama.

Egan, Gabriel. 2006. *Green Shakespeare: From Ecopolitics to Ecocriticism.* London; New York: Routledge.

Emmott, Stephen (Dir. Katie Mitchell). 2012. *Ten Billion.* Performance. Royal Court Theatre, London, July 12 - August 11.

Enelow, Shonni. 2014. Carla and Lewis. In *Research Theatre, Climate Change, and the Ecocide Project: A Casebook*, ed. Una Chaudhuri and Shonni Enelow, 87–116. London; New York: Palgrave Macmillan.

Estok, Simon C. 2011. *Ecocriticism and Shakespeare: Reading Ecophobia*, Literatures, Cultures, and the Environment. London; New York: Palgrave Macmillan.

Extinction Rebellion. n.d. Our Demands. *Extinction Rebellion.* Accessed April 30, 2019 https://rebellion.earth/the-truth/demands/.

Fensham, Rachel, Eddie Paterson, and Paul Rae, eds. 2018. On Climates. *Performance Research* 23 (3): 1–138.

Fevered Sleep. 2010. *The Weather Factory.* Performance/Installation. National Theatre Wales, December 7–24.

———. 2013. *Above Me the Wide Blue Sky.* Performance, Young Vic, London, March 27.

Fried, Larry K., and Theresa J. May. 1994. *Greening up Our Houses: A Guide to a More Ecologically Sound Theatre.* New York: Drama Book Publishers.

Fuchs, Elinor, and Una Chaudhuri, eds. 2002. *Land/Scape/Theater.* Ann Arbor: University of Michigan Press.

Ghelani, Sheila, and Sue Palmer. 2018. *Common Salt.* Performance. London: Battersea Arts Centre, June 7–9.

———. 2020. *Common Salt.* Performance. Reading, UK: Museum of English Rural Life, University of Reading, January 20.

Giannachi, Gabriella, and Nigel Stewart, eds. 2005. *Performing Nature: Explorations in Ecology and the Arts.* Bern: Peter Lang AG.

Glotfelty, Cheryll. 1996. Introduction: Literary Studies in an Age of Environmental Crisis. In *The Ecocriticism Reader: Landmarks in Literary Ecology*, ed. Harold Fromm and Cheryll Glotfelty, xv–xxxv. Athens: University of Georgia Press.

Gray, Nelson, and Sheila Rabillard. 2010. Theatre in an Age of Eco-Crisis. *Canadian Theatre Review* 144 (1): 3–4.

Hamilton, Jennifer Mae. 2017. *This Contentious Storm: An Ecocritical and Performance History of King Lear*, Environmental Cultures Series. London; New York: Bloomsbury Academic.

Hann, Rachel. 2020. Justice Scenographics: Preparing for Civilization Change in a Time of "Anywheres" and "Somewheres". In *Spatial Justice 2.0*, 18–23. Design Studio for Social Intervention, April.

Hansberry, Lorraine. [1959] 2009. *A Raisin in the Sun*. Reprinted. Methuen Drama Modern Classics. London: Methuen.

Harradine, David. 2013. *It's the Skin You're Living In*. Film. Fevered Sleep.

Harvey, David. 1993. The Nature of Environment: Dialectics of Social and Environmental Change. In *Real Problems, False Solutions*, ed. R. Miliband and L. Panitch, 1–51. London: Merlin Press.

Heddon, Deirdre, and Sally Mackey. 2012. Environmentalism, Performance and Applications: Uncertainties and Emancipations. *Research in Drama Education: The Journal of Applied Theatre and Performance* 17 (2): 163–192.

Heim, Wallace. 2012. Epilogue: Thinking Forward. In *Readings in Performance and Ecology*, ed. Wendy Arons and Theresa J. May, 211–216. London; New York: Palgrave Macmillan.

———. 2014. The Continuing Lure of Ecological Performance. In *Landing Stages: Selections from the Ashden Directory of Environment and Performance 2000–2014*, ed. Wallace Heim and Eleanor Margolies, 5–8. London: Crinkle Crankle Press.

Heim, Wallace, and Eleanor Margolies. 2014. *Landing Stages: Selections from the Ashden Directory of Environment and Performance 2000–2014*. London: Crinkle Crankle Press.

Heinlein, Kurt Gerard. 2007. *Green Theatre: Promoting Ecological Preservation and Advancing the Sustainability of Humanity and Nature*. Saarbrücken, Germany: VDM Verlag Dr. Muller.

Heise, Ursula K. 2008. *Sense of Place and Sense of Planet: The Environmental Imagination of the Global*. Oxford; New York: Oxford University Press.

Hickson, Ella. 2016. *Oil*. London: Nick Hern Books.

hooks, bell. 1990. *Yearning: Race, Gender, and Cultural Politics*. Boston: South End Press.

Hopfinger, Sarah. 2018. Intergenerational Performance Ecology: A Practice-Based Approach. *Research in Drama Education: The Journal of Applied Theatre and Performance* 23 (4): 499–516.

Hulme, Mike. 2009. *Why We Disagree About Climate Change: Understanding Controversy, Inaction and Opportunity*. Cambridge: Cambridge University Press.

———. 2010. Why We Disagree about Climate Change. In *The Carbon Yearbook*, 41–43. London: The ENDS Report.

Jacobson, Lynn. 1992. Green Theatre: Confessions of an Eco-Reporter. *American Theater* 8 (11): 17–25.

Joseph, Dione. 2019. Cradling Space: Towards an Indigenous Dramaturgy on Turtle Island. In *Performing Turtle Island: Indigenous Theatre on the World Stage*, ed. Jesse Rae Archibald-Barber, Kathleen Irwin, and Moira Jean Day, 131–145. Regina: University of Regina Press.

Kershaw, Baz. 1992. *The Politics of Performance: Radical Theatre as Cultural Intervention*. London; New York: Routledge.

———. 2002. Ecoactivist Performance: The Environment As Partner in Protest? *TDR/The Drama Review* 46 (1): 118–130.
———. 2007. *Theatre Ecology: Environments and Performance Events*. 1st ed. Cambridge: Cambridge University Press.
Klein, Naomi. 2014. *This Changes Everything: Capitalism vs. the Climate*. London: Allen Lane.
Knowles, Ric. 2015. Devising and Dramaturgy: Decolonizing Praxis. In *Alternative Dramaturgies of the New Millennium in Arabo-Islamic Context and beyond: Selected Papers from the 10th Annual Meeting of the Tangier International Conferences: Performing Tangier 2014*, ed. Khālid Amīn and George F. Roberson, 29–38. Amherst; Denver; Tangier; Mexico City: Collaborative Media International.
Lavery, Carl. 2016. Introduction: Performance and Ecology—What Can Theatre Do? *Green Letters* 20 (3): 229–236.
———, ed. 2018. *Performance and Ecology: What Can Theatre Do?* Abingdon, Oxon; New York: Routledge.
———. 2019. How Does Theatre Think Through Ecology. In *Thinking through Theatre and Performance*, ed. Maaike Bleeker, Adrian Kear, Joe Kelleher, and Heike Roms. London; New York: Methuen Drama.
Lavery, Carl, and Clare Finburgh, eds. 2015. *Rethinking the Theatre of the Absurd: Ecology, the Environment and the Greening of the Modern Stage*. Methuen Drama Engage. London: Bloomsbury Methuen Drama.
Lewis, Simon L., and Mark A. Maslin. 2015. Defining the Anthropocene. *Nature* 519 (7542): 171–180.
Luna, Violeta. 2014. *NK 603: Action for Performer & e-Maiz*. Performance. Massey University, Wellington, June 27.
Macmillan, Duncan. 2011. *Lungs*. London: Oberon Books.
Marranca, Bonnie. 1996. *Ecologies of Theater: Essays at the Century Turning*. Baltimore: The Johns Hopkins University Press.
Masson-Delmotte, V., P. Zhai, H.-O. Pörtner, D. Roberts, J. Skea, P..R. Shukla, A. Pirani, et al. 2018. Global Warming of 1.5°C. An IPCC Special Report on the Impacts of Global Warming of 1.5°C above Pre-Industrial Levels and Related Global Greenhouse Gas Emission Pathways, in the Context of Strengthening the Global Response to the Threat of Climate Change, Sustainable Development, and Efforts to Eradicate Poverty. Geneva, Switzerland: World Meteorological Organization. https://www.ipcc.ch/sr15/.
Matthewman, Sasha, Molly Mullen, and Tamati Patuwai. 2015. The River Talks: An Ecocritical "Kōrero" about Ecological Performance, Community Activism and "Slow Violence". *Research in Drama Education: The Journal of Applied Theatre and Performance* 20 (4): 442–463.
May, Theresa J. 1999. Frontiers: Environmental History, Ecocriticism and the Kentucky Cycle. *Journal of Dramatic Theory and Criticism* IV (1): 159–178.

———. 2005. Greening the Theater: Taking Ecocriticism from Page to Stage. *Journal of Interdisciplinary Studies* 7 (Fall): 84–103.
———. 2007. Beyond Bambi: Toward a Dangerous Ecocriticism in Theatre Studies. *Theatre Topics* 17 (2): 95–110.
———. 2020. Earth Matters on Stage: Ecology and Environment in American Theater. London; New York: Routledge.
May, Theresa J., and Klamath Theatre Project. 2014. Salmon Is Everything. In *Salmon Is Everything: Community-Based Theatre in the Klamath Watershed*, 23–92. Corvallis: Oregon State University Press.
Mendieta, Ana. 1973. *Siluetas Series*.
Metis Arts. 2018. *WE KNOW NOT WHAT WE MAY BE*. Performance. Barbican, London, September 6–9.
Moraga, Cherríe. 1994. *Heroes and Saints and Other Plays*. Albuquerque: West End Press.
Morton, Timothy. 2010. *The Ecological Thought*. Cambridge: Harvard University Press.
Munk, Erika. 1994. Up Front A Beginning and an End. *Theater* 25 (1): 5–6.
Nash, Roderick. 1982. *Wilderness and the American Mind*. New Haven: Yale University Press.
Neal, Lucy. 2015. *Playing for Time: Making Art as If the World Mattered*. London: Oberon Books.
Nixon, Rob. 2011. *Slow Violence and the Environmentalism of the Poor*. Cambridge: Harvard University Press.
O'Malley, Evelyn. 2018. "To Weather a Play": Audiences, Outdoor Shakespeares, and Avant-Garde Nostalgia at The Willow Globe. *Shakespeare Bulletin* 36 (3): 409–427.
Onwueme, Osonye Tess. 2002. *Then She Said It*. San Francisco: African Heritage Press.
Orta, Lucy, and Jorge Orta. 2014. *Symphony for Absent Wildlife*. Performance, Nuit Blanche, Calgary, Canada, September 20.
Osnes, Beth. 2013. *Theatre for Women's Participation in Sustainable Development*. New York: Routledge.
Pitches, Jonathan, and David Shearing, eds. 2019. On Mountains. *Performance Research* 24 (2): 1–130.
Plantable Performance Collective, Green Stage, and Tanja Beer. 2013. *The Trans-Plantable Living Room*. Performance, World Stage Design, Cardiff, September 13–14.
Rankin, Jennifer. 2019. Forget Brexit and Focus on Climate Change, Greta Thunberg Tells EU. *The Guardian*, April 16, sec. Environment. https://www.theguardian.com/environment/2019/apr/16/greta-thunberg-urges-eu-leaders-wake-up-climate-change-school-strike-movement.
Rapley, Chris, and Duncan Macmillian. 2014. *2071*. London: Royal Court, November 6–15.

Richards, Kimberly, Heather Davis-Fisch. 2020. Editorial: Extractivism and Performance. *Canadian Theatre Review* 182: 5–8.

Rimini Protokoll. 2014. *WELT-KLIMAKONFERENZ (World Climate Change Conference)*. Performance. Deutsches Schauspielhaus Theatre, Hamburg.

Ronder, Tanya. 2015. *F*ck the Polar Bears*. London: Nick Hern Books.

Ryan, Courtney B. 2019. #BPCares: Sinking Oil, Spreadable Détournement. *ISLE: Interdisciplinary Studies in Literature and Environment* 26 (4): 882–900.

Scott-Bottoms, Stephen. 2019. The Rise and Fall of Modern Water: From Staging Abstraction to Performing Place. *Theatre Journal* 71 (4): 415–435.

Smith, Linda Tuhiwai. 1999. *Decolonizing Methodologies: Research and Indigenous Peoples*. New York: St Martin's Press.

Spalink, Angenette, and Jonah Winn-Lenetsky. 2020. 'Introduction'. *Performance Research* 25 (2): 1–5.

Standing, Sarah Ann. 2005. From the Redwood Forest. *American Theatre* 22 (2): 65–69.

Stubblefield, Eeo. 1997. *Stilldance with Anna Halprin*. Dance. Accessed November 27, 2018. http://www.stilldance.net/.

Switzer, Sharon. 2013. *#crazyweather*. Carbon 14: Climate is Culture, Royal Ontario Museum, Toronto. https://www.capefarewellfoundation.com/carbon14/crazyweather/, December 29.

Szerszynski, Bronislaw, Wallace Heim, and Claire Waterton, eds. 2004. *Nature Performed: Environment, Culture and Performance*. Hoboken: Wiley-Blackwell.

Thomas, Arden. 2016. Ecodramaturgy. In *Reading Contemporary Performance: Theatricality Across Genres*, 200–201. London; New York: Routledge.

Thompson, Selina. 2019. *Salt*. London: Faber & Faber.

Thorne, Jack, Matt Charman, Moira Buffini, and Penelope Skinner. 2011. *Greenland*. London: Faber & Faber.

Turner, Cathy. 2010. Mis-Guidance and Spatial Planning: Dramaturgies of Public Space. *Contemporary Theatre Review* 20 (2): 149–161.

———. 2015. *Dramaturgy and Architecture—Theatre, Utopia and the Built Environment*. London: Palgrave Macmillan.

Washburn, Anne (Dir. Robert Icke). 2014. *Mr. Burns: A Post-Electric Play*. Performance. Almeida Theatre, London, June 5 - July 26.

Waters, Steve. 2009. *The Contingency Plan*. London: Nick Hern.

Weaver, Deke. 2013. *Wolf*. Performance, Champaign, Illinois, September.

Weintraub, Linda. 2012. *To Life!: Eco Art in Pursuit of a Sustainable Planet*. Berkeley: University of California Press.

Woynarski, Lisa, Adelina Ong, Tanja Beer, Stephanie Beaupark, Jonah Winn-Lenetsky, Rulan Tangen, and Michelle Nicholson-Sanz. 2020. Climate Change and the Decolonized Future of Theatre. *Theatre Research International* 45 (2): 179–207.

CHAPTER 2

Intersectional Ecologies

Salt is a mineral with hidden capital and colonialist connections. Salt, as a reaction between sodium and chlorine (sodium chloride), is an everyday seasoning, a common addition to food everywhere. It is also an essential part of human and animal diets, in moderation. It has 14,000 uses 'including the manufacturing of pharmaceuticals, the melting of ice from winter roads, fertilizing agricultural fields, making soap, softening water, and dying textiles' (Kurlansky 2002: 5). It is part of our biological makeup, 'an adult human being contains about 250 grams of salt' (Kurlansky 6). It has been one of the most sought-after exchange commodities in history: it represented wealth, was used for profit-making, was taxed to raise money for wars and inspired revolutions (Ibid). Either mined underground or harvested from evaporated seawater, the providence of common salt is rooted in complex ecological, social and political contexts. Although now ubiquitous and inexpensive, after the arrival of cultivated agriculture 'salt became one of the first international commodities of trade; its production was one of the first industries and, inevitably, the first state monopoly' (Kurlansky 12). In the seventeenth and eighteenth centuries during British and Dutch colonialism in the Caribbean, salt was the leading cargo transported to North America while salt cod, which fed enslaved people on sugar plantations, was the leading cargo traded in the opposite direction. Salt is tied to colonialism and slavery (slave labour was also used in salt production in the Caribbean) as well as ecological destruction as trees were clear cut to fuel salt pans on some Caribbean islands, which degraded

L. Woynarski, *Ecodramaturgies*, New Dramaturgies,
https://doi.org/10.1007/978-3-030-55853-6_2

the soil and freshwater supply (211–213). Salt in India during the rule of the British East India Company (1600–1858) and then the British Raj (1858–1947) was exploited through extortionate taxation for British wealth accumulation at the expense of Indian people, which eventually led to Gandhi's famous Salt March and Indian independence. The multiple stories embedded in salt reveal the ways in which ecological history and human history are one and the same, shaped by intricate entanglements of power relations and materiality. As a material, salt contains and represents these stories of ecological inequality. The salt tax in India, and the hedgerow used by the British to enforce it,[1] was the impetus for *Common Salt* (2018–2020), a show-and-tell performance by Sheila Ghelani and Sue Palmer. Salt stands in for Europe and a preservative for the legacy of the colonial and imperial violence that shaped transatlantic slavery in *Salt.* (2017–2019), a solo performance by Selina Thompson. *Then She Said It* (2002) by Osonye Tess Onwueme, about rural women protesting oil extraction in the Niger Delta, centres on oil as another global commodity replete with colonial violence. The Green Belt Movement of tree planting in Kenya reveals the connections between colonial power structure, land management and gender roles. Through their unravelling of complicated ecological, colonial and political-economic stories of salt, hedges, mazes, corporations, slavery, labour, petrol, trees and trauma, these works perform intersectional ecologies.

The interlocking structure of salt, and its indexical meanings based on its history as a commodity, helps to address one of the central concerns of this book: how ecodramaturgies can expose the series of interconnected ecological problems and injustices that are often hidden, erased or simplified. Ecological issues are often presented in a homogenous way, based on a narrative of a singular problem in need of a 'solution'. The issues are regularly whitewashed and patriarchal, shaped in relation to certain bodies and identities (Finney 2014). Ecological relationships are frequently invisible, particularly in consumer relations. It is easy to buy something but much more difficult to find out who made it, how it was made, what resources and materials were used and what happens when it is thrown away. Cultural works, such as theatre and performance, can reproduce these problems by creating an image of ecological work as 'green and pleasant', middle class, white, singular and reductive.[2] In order to address this, an intersectional understanding of ecology is needed to expose the complexity and interlinked issues of oppressions and injustices. Towards this aim, salt becomes a useful metaphor and analogue for revealing the way in which seemingly normalised relations and objects are concealing complex histories of colonial and capitalist logics that produce unequal effects. The performances discussed in this chapter—the Green Belt Movement, *Common Salt*, *Salt.* and *Then She Said It*—can all be read as performing intersectional ecologies by revealing the density of interrelated issues that create ecological injustices.

'Intersectionality' was coined by legal scholar Kimberlé Crenshaw (1989) to address complex situations of injustice and inequality. Crenshaw used the term to grapple with the overlap of race and gender in the US legal system in response to a 'single-axis framework' of discrimination. This had a 'tendency to treat race and gender as mutually exclusive categories of experience and analysis' and consequently erased the experience and visibility of Black women (Crenshaw 1989: 139). Crenshaw advocated for an intersectional framework to account for the experiences of Black women, seeing it as a much-needed addition to address a history of exclusion in Western feminist theory and anti-racist politics. It has since been extended to multiple experiences and identities, based on the principle that human lives and socially constructed/enforced identities are complex, layered, relational, contextually contingent and fluctuating. Scholar Patricia Hill Collins writes that 'Intersectional paradigms remind us that oppression cannot be reduced to one fundamental type, and that oppressions work together in producing injustice' (Collins 2000: 180). I employ intersectionality as a way of reading and as praxis, as Luft and Ward describe: 'a practice and a political intervention' (2009: 10). I resist a homogenising theory of intersectionality and use it as a way of thinking about the power structures that shape ecological relationships and their ideological underpinning. As praxis, I see it as a commitment to work against the notion of ecological concerns being white, middle-class problems, which can be fixed through 'conscious' capitalism (such as buying 'eco-friendly' consumer goods). This means centring voices of marginalised people, considering who (human and more-than-human) has the social capital to resist systematic ecocide. Intersectional ecologies analyses the ideologies of conquest, colonialism, accumulation and individualism that shape capitalist patriarchy and ecological exploitation.

Critiques of intersectionality include its status as a vague 'buzzword', the fear that it is the latest feminist cliché that will eventually become, as scholar Ange-Marie Hancock writes, 'ubiquitous in its familiarity but devoid of tangible political impact' (Hancock 2016: 7). Feminist scholar Kathy Davis describes intersectionality as open-ended and ambiguous: 'It is not at all clear whether intersectionality should be limited to understanding individual experiences, to theorising identity, or whether it should be taken as a property of social structures and cultural discourses' (Davis 2008: 68). Scholar María Lugones offers a linguistic critique of intersectionality, arguing that the language of 'intersecting' or 'interlocking' identities of race and gender follows a hegemonic logic which can reassert rather than dismantle these oppressions or that 'the intersecting hides the inseparability of oppressions' (2014: 76). The experiences of oppression cannot be separated, even for the purposes of conceptual analysis. She argues that this can lead to fragmentation of self-identity. What is at stake for Lugones is the loss of potential to resist the logics and structures of oppression when using the very language of hegemonic discourse that categorises and separates. She therefore advocates for a 'logic of fusion' using terms such as 'intermeshing' to resist the logics of white supremacist heteropatriarchy. Decolonial feminist Emma Velez suggests that fusion and intersectionality are complementary, despite Lugones' critique, as an analysis is needed of both the power structures that shape the categories of oppression as intersecting and the experiences of the reality of these fusions (2019).

Here the relationship between decolonising and intersectionality becomes potent as it aims to unravel 'how coloniality buttresses the oppressive categorial logics that intersectionality identifies' (Velez 2019: 392). Lugones' coloniality of gender 'is what lies at the intersection of gender and class and race as central constructs of the capitalist world system of power' (2012: 75). The analysis of these intersections aims to understand how the processes of colonisation enact a complicated layering of dehumanisation, racialisation and gendering. Lugones argues that just as colonised people were not meant to occupy the category of 'human', colonised women were excluded from the category of 'woman'. Decolonial feminisms are about tracing and then ending the oppressions of colonialism by building coalitions. The performances in this chapter—the Green Belt Movement, *Common Salt*, *Salt.* and *Then She Said It*—assert the first aim of tracing the processes of colonisation that cause intersecting oppressions. By exposing the effects of the logics of colonialism that result in violent greed, exploitation, slavery and death, undergirded by capitalist systems, the performances question these logics. Through revealing how colonialism created poverty, death, human chattel slavery, disenfranchised women and environmental destruction, the performances push against the categorisation of gender, nationality, race and class.

Intersectionality presents a number of questions to ask about how power, identity and agency interact to create complex differences and oppressions. Legal scholar Mari Matsuda conceptualises it as an approach of 'asking the other question': 'when I see something that looks racist, I ask "Where is the patriarchy in this?" When I see something that looks sexist, I ask "Where is the heterosexist in this?" When I see something that looks homophobic, I ask "Where are the class interests in this?" (1991: 1189). Intersectionality is a tool for a multilevel analysis of how oppressive structures work in relation to each other. Researchers Kaijser and Kronsell offer a useful set of starting questions for thinking about the intersectionality of climate change: 'How are relations between humans and between humans and the environment portrayed? How is nature represented? What type of environmental knowledge is recognised and privileged?' (2014: 430). These questions also apply to ecodramaturgies in my analysis, in addition to: Who is being represented and who is being erased in the narrative? What ideology, knowledge or subject position is represented as the 'norm' or dominant? What places and people are bearing the brunt of ecological consequences? How is the more-than-human being represented? How is the slow violence of climate change, and other ecological violence, producing inequality and fatal, asymmetrical effects? Unravelling these questions and naming these practices are what I call intersectional ecological thinking.

The focus of this chapter is how ecodramaturgies can make visible the topographies of intersectional ecologies. Distilling the multiple indexical relationships in an abstracted commodity such as salt, as well as its intrinsic biopolitical connectivity, requires multilevel analysis of intersecting power relations that organise the lived experience of fusion and the current economic moment. Intersectional ecologies are about recognising that ecological issues, such as climate change, cannot be thought of in isolation or as single issues. They demand an understanding of the political forces and ideologies that shape ecological relationships. On a global scale, ecological destruction disproportionally affects women, the poor and people of colour through slow violence and is therefore tied to other forms of violence, marginalisation, political power and social mobility (Nixon 2011). Intersectional ecologies attend to how identities and power relations construct and regulate ecological responsibility and response-ability, foregrounding the way ecological issues are multi-sited and also always social, political, gendered, racialised, class-based, disability and access issues. Ecodramaturgies expose the way in which ecologies are intersectional.

Theatre and performance stories, images and narratives can lay bare the interrelated structures and power relations and their ideological foundations that create injustice across multiple axes. Ecodramaturgies can counter the abstraction and reduction of these issues by locating them in potent stories, missing histories, forgotten people and places, and human and more-than-human bodies. The discussion in this chapter unfolds in four parts beginning with the intersectional ecofeminisms in women-led environmental movements in the short play *It Starts With Me* (2019), and then focusing on the Green Belt Movement led by Wangari Maathai. It then moves to the object ecodramaturgy of recuperating histories and presents of colonialism in *Common Salt* and salt as the ongoing legacy of slavery, capitalism and racism in *Salt*. The ideologies of these legacies repeat themselves in oil production in *Then She Said It*.

Women-Led Movements

Intersectional ecologies come from a legacy of ecofeminism and women-led environmental movements. The unequal effects and burden of climate crisis is the contemporary iteration of the oppressive logics of capitalist, neocolonial heteropatriarchy. The short play, *It Starts With Me* (2019) by US/Canadian playwright Chantal Bilodeau (part of Climate Change Theatre Action 2019) focuses on the way in which climate crisis disproportionately affects women. Described as a battle cry, the piece is performed with several different women who trade dialogue and build up a chorus of voices. The uneven effects of climate change, along gender lines, are articulated energetically in the play, encouraging identification and solidarity with those in the audience who identify as women. Single voices start by identifying themselves as 'sister', 'mother', 'niece' and then 'student', 'scientist', 'environmentalist', 'an Elder'. The performers then declare some of the action they can take despite not having a lot of power, such as despite being young 'I can stand in front of world leaders, and demand that they do their job' or despite being marginalised 'I can draw upon indigenous traditional knowledge, and heal the Earth' (Bilodeau 2019: 2). The energy and intensity begin to build as statistics about the gendered effects of climate change come in:

> SINGLE VOICE: Because I'm 14 times more likely to die in a natural disaster
> Than a man!

SINGLE VOICE: Because I'm 4 times more likely to be a climate refugee
Than a man!
SINGLE VOICE: Because after a natural disaster, I am 300 times more likely to be abused
By a man!
SINGLE VOICE: Because I'm vulnerable to toxins and pollutants!
MULTIPLES VOICES: Always!
SINGLE VOICE: To human trafficking!
MULTIPLES VOICES: Always!
(Bilodeau 2019: 3)[3]

The play builds to a crescendo, with the actors stomping their feet proclaiming, 'it starts with me'. The stage directions read 'channel all the emotions—all the hope, rage, fear, disappointment—you have about being a woman in this world' (Bilodeau 2019: 4). The performers encourage all those who identify as women in the audience to join in, repeating 'it starts with me' until they do. After the crescendo of stomping and chanting, the performers hug each other and the women in the audience, if appropriate, signalling the solidarity they are aiming to build.

The play lists some of the ways in which women are more vulnerable to the effects of climate change and how they are standing up. Wangari Maathai, leader of the Kenyan Green Belt Movement, was one of the inspirations for the play. The dedication reads: 'this play is inspired by Greta Thunberg, Katharine Hayhoe, Wangari Maathai, Alexandria Villaseñor, Naomi Klein, Rebecca Solnit, Sheila Watt-Cloutier, Alexandra Ocasio-Cortez, Christiana Figueres, and countless more women who are fighting for us all' (Bilodeau 2019: 1). The inclusion of statistics on climate change refugees and violence is designed to connect women across geographical locations, races, ages and classes. This solidarity building between women, who are at the forefront of climate change effects and the extractivist ideology behind it, are present in the women-led Green Belt Movement.

Green Belt Movement

The story of Kenya's Green Belt Movement is a compelling narrative of a historic and effective women-led movement whose success may be attributed to strategies of intersectional environmentalism through a dramaturgy of tree planting. Wangari Maathai, founder of the movement, addressed both environmental and socio-political security, which are intimately intertwined. Nixon notes that poverty and environmental conditions are so connected that any strategy for addressing poverty needs to be environmental (2011: 129). To combat the slow violence of soil erosion in Kenya that disproportionally affected women, as they are the principal drivers of subsistence agriculture, Maathai created a large-scale tree planting movement, beginning in 1977, with more than 6000 local tree nurseries, employing 100,000 women and planting 30 million trees (Nixon 2011: 129). People in Kenya had moved their land to cash crops such as coffee and tea which resulted in a loss of trees and firewood for cooking traditional food crops as well as a lack of water. This in turn changed food habits to highly refined food, rich in carbohydrates and low in protein and vitamins (Maathai in Dater and Merton 2008). When Maathai began the project, she needed to teach women how to plant trees and see the importance of this by making connections between environmental problems and the problems of their daily lives. The women were taught to collect seeds from trees in their area and propagate them. They were then compensated with a small amount of money for every tree that survived.

Maathai's research into the environmental problems of Kenya revealed that they were the symptoms of wider political systems. British colonialism in the late 1800s stole the best land, cut down forests and displaced the Indigenous peoples to impose European ways of farming, religion and governing. The British violently suppressed those who resisted colonial rule (such as the Mau Mau) through massacres, bombings, burning of homes and lands including forests where they resided (Dater and Merton 2008). Tea and coffee replaced Indigenous food crops. The colonial practice of deforestation continued in the Kenyan independent governments from 1963. The neocolonial governments, ruled by Kenyan elites, viewed the natural resources of the land as property of the government (just as they were taught by the colonialists to think of them as 'crown lands'). Opposition was crushed through imprisoning dissentients and activists. Maathai viewed the relatively simple act of tree planting as also planting ideas, empowering women to stand up for their rights. The dramaturgy of tree planting, the defining and democratic act of the movement, was as Nixon describes 'a symbolic hub for political resistance' and a 'pragmatic, yet powerfully figurative act that connected with many women's quotidian lives as tillers of the soil' (Nixon 2011: 132–133). For Maathai, work on environmental rights could not happen without first empowering women and communities.

Nixon argues that by aligning the Green Belt Movement with ‘other civil rights campaigns that were not expressly environmental, like the campaigns for women’s rights, for the release of political prisoners, and for greater political transparency’ (Nixon 2011: 133), an intersectional environmentalism emerged that broadened their base of support. Tree planting, or theatre of the tree, acted as a metaphor, an affective material and vibrant symbol of the aims of the Green Belt Movement, tying peace, democracy and civil rights to the ecology of Kenya. However, some forestry initiatives for development schemes have been criticised, as Green et al. detail, for ‘appropriating women’s labour, unremunerated’ or adding additional ‘chores’ to their caring responsibilities or treating women as ‘cheap labour’ for environmental projects of which they are not always involved in the decision making (Green et al. 1998: 274–275). The women in the Green Belt Movement, however, had agency, and with Maathai at the helm they were able to make decisions about what was best for them. Maathai considered tree planting a way of breaking the cycle of poverty, saying ‘poverty is both a cause and a symptom of environmental degradation’ (Maathai in Nixon 2011: 137). The Green Belt Movement is an example of intersectional ecological thinking because of the way in which multiple perspectives and experiences were attended to, acknowledging the interconnections between the overlapping challenges facing Kenyans in relation to gender, democracy, civil liberties and food access, all of which are ecological.

The dramaturgy of tree planting, or the theatre of trees as Nixon describes it, is an act that is relatively easy to gain popular support for. The act of planting and growing is non-violent and for the benefit of many. For Nixon, ‘The theatre of the tree afforded the social movement a rich symbolic vocabulary that helped extend its civic reach. Maathai recast the simple gesture of digging a hole and putting a sapling in it as a way of “planting the seeds of peace”’ (Nixon 2011: 133). At first, it went under the radar of the Kenyan government: ‘When the women started nobody was bothering them because nobody took them seriously, because who took women seriously’ (Maathai in Dater and Merton 2008). It then gained momentum, becoming powerful because it became seen as representing a threat to the power structures of Kenya, led by women. Tree planting came to be considered an act of civil disobedience by the Kenyan authorities. The Green Belt Movement has continued after Maathai’s death in 2011 and has expanded its work to include rehabilitation of watersheds, tackling the effects of climate change, building a network of women climate leaders, and gender livelihood and advocacy. Their stated mission 'is to strive for better environmental management, community empowerment, and livelihood improvement using tree-planting as an entry point' (Green Belt Movement 2016). Tree planting is a powerful protest and mode of empowerment at a grassroots level.

Maathai's fight for the environment was also a fight for democracy and for gender equality. She was the first Kenyan woman to earn a PhD and then a university chair and the first African woman to win the Nobel Prize (2004). She divorced her husband when divorce was still taboo in Kenya. Kenyan playwright MaryAnn Karanja imagines a meeting between a Maathai avatar and a young woman who is a sex worker in a jail cell, in the short play *Birthday Suit* (2019) part of Climate Change Theatre Action 2019. The young woman, Pam, is inspired by the older environmental activist (called Riziki in the play) who is imprisoned for stripping outside the state house in protest of the deforestation of Karura Forest to make room for development. Riziki says 'I undressed in protest. To make a statement to the government that I'm ready to do anything to ensure that deforestation and grabbing of Karura forest stops' (Karanja 2019: 2). Pam slowly begins to identify with Riziki despite their differences in age, class, education and career, understanding how they are both in oppressive structures. 'It's funny how different we seem yet we are so alike! You strip, I strip. You strip for trees, I strip for men; the only difference is, men pay instantly, trees pay later' (Karanja 2019: 4). The two women in the play are both imprisoned for gender disobedience within a system of capitalist patriarchy. The naked protests were used by some of the women in the Green Belt Movement to resist the violence of police in a fight to release political prisoners. The nudity of older women is considered a curse in Kenya and the only form of resistance or self-defense the Green Belt women felt they had against the violence of they were facing (Dater and Merton 2008).

As alluded to in *Birthday Suit*, the conflict over the clearing of Karura Forest was a dramatic standoff between the neocolonial Moi government and the Green Belt Movement beginning in 1998. This standoff typified the conflicting ideas of land and environment held by the government and Maathai. Moi felt the forest belonged to the state and therefore the government ministers were entitled to do with it what they liked, such as sell it off for development of luxury gated housing, producing economic 'growth'. Led by Maathai, the Green Belt Movement viewed the forest as public land, belonging to the commonality, and its deforestation was a 'last straw' of an undemocratic government that disenfranchised the poor to benefit the rich. The rights of the poor were being violated. The Green Belt Movement protestors, armed with tree saplings, stood off against the government-funded armed guards who used violence to suppress them. The image of a tree sapling against armed guards was a powerful symbol that acted as a catalyst for many other marginalised people to join anti-government protests. Maathai sustained a bloody head wound and the protestors were imprisoned (Nixon 2011: 135). After a year of protests and violent clashing with the police, the development in Karura Forest was stopped. The Moi government was defeated in 2002 after 24 years in power and Maathai was elected to parliament (2002–2007). The new government supported reforestation with the military planting one million trees.

The Green Belt Movement can be read as part of a legacy of ecofeminists from the Global South pioneering intersectional approaches in their activism. Maria Mies and Vandana Shiva write about an ecofeminist approach to address the capitalist patriarchal world system which 'is built upon and maintains itself through the colonisation of women, of "foreign" peoples and their lands; and of nature, which it is gradually destroying' (1993: 2). Mies and Shiva assert that women-led environmental movements all over the world have demonstrated the shared feelings of anxiety and anger coupled with a sense of responsibility despite their differences: 'irrespective of different racial, ethnic, cultural or class backgrounds, this common concern brought women together to forge links in solidarity with other women, people and even nations' (1993: 3). The solidarity of activist environmental movements led by women characterises the ecofeminism Mies and Shiva detail, including campaigns against mega dams on the Narmada River in India. They describe some early intersectional ecological practices that centre perspectives from the Global South, foregrounding capitalism and colonialism in their marginalisation as well as the more-than-human world. The performance piece *Common Salt* uses objects to tell stories of British colonialism in India and how capitalist mechanisms are intimately tied to colonial power structures.

Common Salt

Common Salt (2018–2020) by Sheila Ghelani and Sue Palmer tells multiple stories about the ongoing effects of British colonialism in India and some of its ecological implications. The stories told in performance include the Great Hedge of India, which was a large-scale hedgerow used as a

Customs Line by the East India Company to enforce the salt tax in India; the way colonialism shapes trade and money today; the maze at Hampton Court Palace; the land Enclosures Act in Britain; nineteenth-century naturalist Eliza Brightwen and her home museum in Stanmore, London, which later became the site of Marconi Defence Systems; and the artists' journey of researching and making the performance. The artists describe it as 'an expansive and emotional time-travel' from the seventeenth to the twenty-first century, covering 'narratives of trade, race and culture' (Ghelani and Palmer 2020b). My account of the performance comes from a showing at the Battersea Arts Centre, London in 2018, and the Museum of English Rural Life (MERL) at the University of Reading in 2020. It has toured to other museums, galleries, libraries and art spaces. The show-and-tell performance resonates with the museum location because it is about how the stories of objects and histories are told, erased and forgotten.

Common Salt, performed by Ghelani and Palmer, has an audience of about 25 people. The whole performance takes place around the 'nature table', which is covered in a grey fringed tablecloth and set with an easel and a sign board reading 'Common Salt: A show and tell', a Shruti box and two rolled up neckties (Fig. 2.1). The audience sits around three sides of the table with performers Ghelani and Palmer standing on the fourth side. There is no theatrical lighting, just a balanced-arm task lamp which is moved around as a spotlight to illuminate the objects under discussion. Ghelani and Palmer are dressed in vibrantly printed shirt dresses: bright green hedges for Palmer and yellow with cartoon tiger heads for Ghelani. They both have the pages of the script in front of them. We are welcomed by the two performers as we enter and settle into the space. There is a sense of intimacy for the small audience as they make eye contact with us, although we are told not to worry there is no audience participation. Bird song is playing through a small set of speakers connected to an iPhone. The performers introduce themselves and the 'show and tell' structure is explained. The piece is conversational although the audience is not asked to respond.

The show has a deceptively simple setup: a collection of objects presented on the nature table, which the performers describe and tell stories about. It utilizes a choreography of everyday objects, producing an eco-dramaturgy that connects the everyday objects to ecological stories, past and present. At different points, placed on the table are 'nature' objects (moss, bird's nest, flowers, plastic balls of boxwood, scallop shells, model trees), capitalist and money objects (casino chips, croupier stick, £20

Fig. 2.1 Sheila Ghelani and Sue Palmer, *Common Salt* at the Queen's House, Royal Museums Greenwich in front of the Armada Portrait of Queen Elizabeth I, 2019. (Photo by: John Hunter, courtesy of the artists)

notes, scales) and colonialist objects (a painting of an East India Company ship, mini British military figures, a china saucer painted with cranes). The objects are catalysts for multiple, interwoven stories that explore the history of British colonialism in India, militarism and the salt tax, the East India Company and neoliberal capitalism, hedgerows, hedge mazes and hedge funds, and the performers' own experiences creating the performance. The choreography of objects is playful yet depicts the violence of colonialism. Objects are placed purposefully on the nature table, sometimes with a flourish and sometimes with a litter picker or moved around the table with a croupier stick representing the capital and colonial forces shaping ecological relations. By the end of the performance, the nature table is full of objects of the stories told.

The promotional material for *Common Salt* describes it as 'a reckoning; the interconnectedness between history, empire, nature and memory is hidden in plain sight' (Ghelani and Palmer 2020b). This reckoning is intersectional as it takes the abstract commodity of common salt and exposes some of its biopolitical and indexical relations through stories of objects. The ecodramaturgy follows the 'logics of fusion' (Lugones 2014) as the stories are intermeshed to make clear that ecology is not a single issue, environmental history is human history and there are many overlapping and differentiated social and political forces that shape ecological relationships. Of the many stories contained in *Common Salt,* my analysis follows colonialism, the East India Company and the residue of how these ideologies continue to shape ecological relations. Countering the 'violence of abstraction' (Hartman 2008: 7), the performance references the capitalist logics that underpinned the activity of the East India Company, which were structured by colonial constructs of identity that dehumanised Indian people and commodified natural resources.

The piece is structured as 'seven turns and an island' which was inspired by a visit to the hedge maze at Hampton Court Palace. The artists were relatively unimpressed with the hedge maze in Hampton Court Palace (hedges were bare and easy to see through, in the middle of the maze was a corporate sign) and more intrigued by the display of guns and military paraphernalia inside the Palace. They connect that visit both to the historical hedgerow planted in India as a Customs Line and the British imperialist, military-supported colonisation of India, which lined the pockets of wealthy Britons. Laments are sung by the performers as a form of introduction and narration. Ghelani plays the Shruti box, a wooden instrument that creates tonal sounds based on the expansion and contraction of forced air through the box, used in Indian classical music, while Palmer accompanies her on a small pitch pipe. These laments are self-reflective and full of humour and playfulness: 'We went to see the maze at Hampton Court Palace, but it was a bit disappointing...And in the centre there was an awful silver sign. Nothing wonderful. Nothing extraordinary. Just an awful silver *corporate sign*' (Ghelani and Palmer 2020a). The tonal drones of the Shruti box provide an eerie atmosphere for detailing the, at times violent and at times surprising, stories. They tell us it started 400 years ago when the East India Company was given a Royal Charter by Elizabeth I in 1600, able to trade commodities, 'mint its own money and raise an army. Here is the seed of the British Empire' (Ghelani and Palmer 2020a). The piece makes connections visible by situating it geographically in London (with some details changing based on where it is being performed) as well as personally, as Ghelani is of Indian/English heritage and spoke to her Indian-born father as part of the research for the piece (he did not know anything about the East India Company), and Palmer is White British whose Great Aunt came from a colonial family with interests in India.

During the performance, Palmer holds up a china saucer with painted cranes on it that belonged to her Great Aunt as she explains 'The family ran tea plantations in India, and security at a gold mine. And on occasion, hunted big game for sport—tigers' (Ghelani and Palmer 2020a). The non-linear narrative of the piece tells the performers' own journey of researching and recovering forgotten stories about ecology, collecting, colonialism and neocolonialism.

The fourth turn involved recounting a journey on the Jubilee line of the London Underground from Stanmore to Canary Wharf, connecting the colonial salt tax to hedge funds and modern neoliberal capitalism 'from Hedgerows to hedge funds—then to now…Money and the past. Money and the present. Cash and colonialism. Cash and collecting' (Ghelani and Palmer 2020a). The narrative device of the tube journey creates a contemporary connection, with the performers at one point opening the shutters in the performance space at the BAC performance to point to the Lloyd's building in the heart of the City of London (the financial district), which is an insurance company located on the site of the former East India Company headquarters, locating the colonial legacy in institutions still present and drawing connections to wealth accumulation.

These stories are not located in the past, rather they demonstrate how the past informs the present with neocolonial power structures still shaping ecological relations. Ghelani and Palmer sing (and discuss) how personal objects now such as tea, takeaways and British passports contain colonial and ecological residues:

> One day this will be forgotten, Like the Great Hedge of India and Eliza Brightwen
> And the East of India Company with all their looting and killing and providing of a model for corporations nowadays.
> Did we learn nothing? It seems as though people learn nothing. That everything is forgotten, that people learn nothing…
>
> When we learn facts about the way humans are treating each other we feel confused…
> And animals and plants and their place in the world. What is their place in this world?
> It seems no place in the world isn't full of foliage and screens, full of Salt and of blood. What about the hedgerows?…

> What's this got to do with us?
> The tea we drink, £1 Tshirts, takeaways, the books with their missing chapters, deportation centres, our British passports. (Ghelani and Palmer 2020a)

Colonial and neocolonial logics still construct the '£1 Tshirts' made by low-waged workers without safety protections in the Global South (often former colonial nations) where there are little environmental protections for manufacturing. In a decolonised tour of the Museum of English Rural Life that preceded the 2020 performance, a Ladybird book depicts how tea was stolen from China by the British East India Company and brought to India to develop the Indian tea industry: 'that English mainstay—the cup of tea—conceals a complex and conflicted colonial history' (Museum of English Rural Life 2020). These colonial logics also shape the nationalist discourse, privileges of travelling with a British passport and the deportation centres in which migration and immigration becomes a scapegoat for underfunded social services and unemployment rates. Who gets to migrate is organised around oppressive categories of nationality, race, ethnicity, class and religion. The British Conservative government has created a 'hostile environment' towards immigration, conveniently forgetting histories of migration and immigration under British colonialism (the Windrush scandal being a prime example[4]). The songs and stories take everyday objects of contemporary life in the UK and dismantle their abstraction by making visible the ecological and social injustices and the ongoing effects of colonialism that are intimately ingrained in them.

The show-and-tell form of the performance utilises a dramaturgy of a collections talk, borrowing conventions from museums, while also deconstructing the way history is communicated. For Ghelani, all of the intermeshed stories of the piece speak to 'the practice of collecting/collections in relation to museums and the "history of forgetting"' (Ghelani in Ghelani and Woynarski 2018). The story of the Salt Tax in India is one of the many forgotten or wilfully ignored aspects of the violences of British colonialism. As Ghelani and Palmer detail in the performance, 'The Salt Tax was born out of British greed… The greed of the East India Company, its servants and its shareholders, and then the British government, its parliament, its electors' (quoted from Moxham 2002: 33). Although everyone in India needed access to salt as an essential part of human and animal diets, 'The Salt Tax created severe hardship—the price of salt needed for a family was worth up to two months wages for an average agricultural labourer' (2020a). They describe the British enforcement of the tax as ruthless, killing those disobeyed and collecting obscene amounts of money for the British aristocracy who held shares in the East India Company. Through a dramaturgy of objects, this is detailed strikingly in the performance by a simple image: Ghelani slowly pours table salt over an antique silver salt cellar held in a silk-lined box. A mountain of salt is created spilling forth from the box, a coin from the East India Company is placed on top of the pile and then two mini figures of British Raj soldiers (one on horseback and one holding a rifle) are placed either side, guarding the mountain of salt (Fig. 2.2). This story addresses the hidden and abstract narrative of colonialism exemplified by a 2016 YouGov poll that revealed 44% of Britons were proud of the British Empire (Singh 2016), indicating the erasure of exploitation, violence, genocide and slavery. By revealing what is conventionally hidden, the image animates some of the colonial and neocolonial aspects of ecological relations.

Later in the performance, Ghelani pulls out the salt cellar from the mountain of salt and places a glass bell jar over it, representing the British

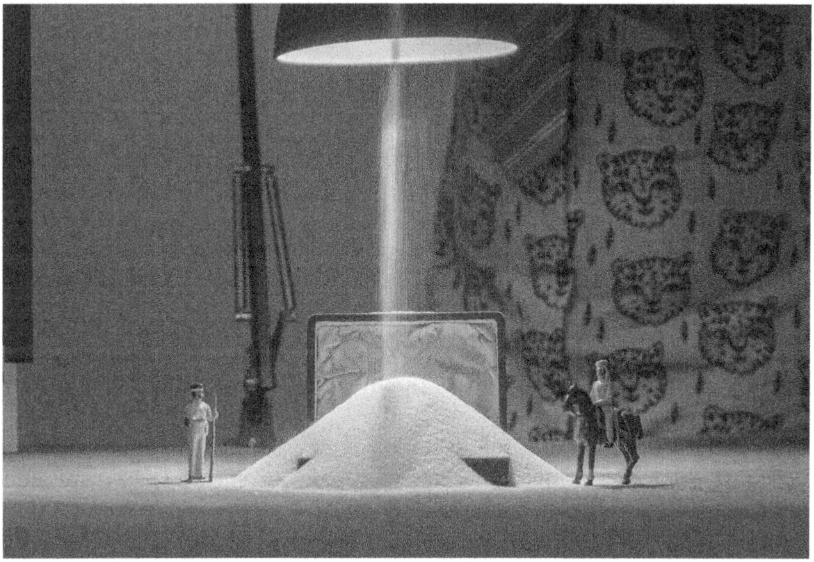

Fig. 2.2 *Common Salt* by Sheila Ghelani and Sue Palmer, Museum of English Rural Life (MERL), Reading, 2020. (Photo: Amie Cliffen, courtesy of Sheila Ghelani and Sue Palmer)

Empire: 'it's behind glass, and it's not labelled' (Ghelani and Palmer 2020a) referencing its abstraction. British colonialism in India was ecologically destructive to landscape and people, in the name of maximising profits for wealthy Britons and the Crown. As Kurlansky details, 'before the British created artificial trade barriers, India had affordable, readily available salt' as it had 'natural salt fields on both coasts and huge rock salt deposits and salt lakes in between, India had an ancient tradition of salt making and trading' (Kurlansky 2002: 334). Before British colonialism, the salt from the Orissa area was sought after and traded extensively and provided a living for the poorest people of the area. After the British took over, they banned Orissa salt to keep British salt merchants happy, and the British Army occupied the area and annexed it to Bengal in order to have a monopoly (Kurlansky 337). The salt-production ban led to famine in the area, which was a precursor to the larger famine in 1876 which killed six and a half million people. Public organising against the salt tax began in 1888 in Orissa, in which it was pointed out that 'impoverished Indians had a tax burden thirty times greater than did people in England' (Kurlansky 342). Salt was not made affordable until Indian independence in 1947. Now Gujarat is India's largest salt-producing area, which employs thousands of extremely low-paid migrant workers who are often undeclared to circumvent minimum wage and labour laws. Working on the salt pans has adverse health effects with many 'workers permanently colorblind', while upon their death, 'their bodies cannot be properly cremated because they are impregnated with salt' (Kurlansky 354). The capitalist logics of exploitation continue while India is the third biggest salt producer in the world. The container of table salt is an everyday object that is fused with a myriad of connections and illustrates the intersections of ecology, colonialism, imperialism, geography, sea, trade, labour and borders within the grand narrative of ecological exploitation of both humans and more-than-humans. *Common Salt* opens our eyes to some of these intermeshed relationships and legacies.

Through the show-and-tell dramaturgy of *Common Salt*, the colonial, capitalist ideologies of the East India Company are shown to be reiterated by neoliberal capitalism, making live connections between the spatial and material links and the structures present today. The East India Company 'provided a model for corporations nowadays' (Ghelani and Palmer 2020a) and is linked to the trading port of Canary Wharf on West India Dock in London, where goods were imported from British colonies. Today, the area is home to multinational corporations and financial services. Talking about their journey, Palmer says 'We found things trimmed, cut, ties being cut, clear cut, felled, sold off' (Ghelani and Palmer 2020a). At this moment, Ghelani brings out a neat stack of casino chips held between the prongs of the litter picker and releases them on the table with a dramatic thud. The violence of colonialism that clear cuts hedges, fells forests and sells off land and natural resources is still ongoing today. The same ideologies are reproducing the hardship felt by the salt tax, the enclosure of common land and the commodification of 'nature' for the benefit of a few, perhaps replaced by the asymmetrical effects of climate change felt in the Global South (particularly in India and Bangladesh). By weaving together the stories of their own journey of making the piece, the hedgerow Customs Line in India and the colonial and neocolonial forces shaping ecological relations, intersectional ecologies are performed.

Common Salt performs stories, histories and present-day accounts of what governing logics have erased and abstracted. The reiteration of colonial power dynamics through the time-travelling stories of salt expose the normalisation of physical and ecological violence embedded in everyday objects. The compelling ecodramaturgy of the performance uncovers uncomfortable relationships and complicity in the ongoing colonial power relations that reinforce oppressive categorisations. In this way, *Common Salt* is a lively 'alternative' history to the popular British myths of nostalgia and pride for the British Empire.

Salt.

Salt. (2017–2019) is a solo performed show written and performed by Selina Thompson in 2017–2018 and then performed by Rochelle Rose during the 2019 run at the Royal Court Theatre, London (as Thompson found there was no way of preserving her well-being through repeated performances). The performance recounts a journey Thompson undertook in 2016, retracing a route of the transatlantic slave trade triangle: leaving Britain, taking a cargo ship from Belgium to Ghana, flying to Jamaica and finally travelling back to the UK. Thompson is a Black British performance maker. This situated knowledge is important to the biographical nature of the piece. Among the first lines of the performance are 'I am twenty-eight. I am Black. I am a woman. I grew up in Birmingham which is where all my family live' (Thompson 2019: 14). She also tells us her grandparents are from Jamaica and Montserrat, and this journey was her first time using her adult passport. The performance covers an array of interconnected stories, experiences and themes; it is an attempt to commemorate those who did not survive, to confront the inherited trauma and ongoing legacy of slavery, to search for home and to grieve. I understand this to be what Christina Sharpe refers to as 'wake work': 'to be *in* the wake is to occupy and to be occupied by the continuous and changing present of slavery's as yet unresolved unfolding' (Sharpe 2016: 13–14). It is also about how this burden shapes the experiences of Black British people today, how the ideologies of colonialism and imperialism are still felt through oppressive social structures. These forces are manifested in the performance through salt. Salt production in the Caribbean and the US relied in slavery at one time. I read this piece as enacting intersectional ecodramaturgies through the unravelling of how these structures are intermeshed and ongoing. Ecology pervades the dramaturgy, implying that colonialism, slavery and Black British identity are shaped by these forces which are also ecological.

The smell of incense fills the air, while on stage is an altar, a sledgehammer, safety goggles and gloves. A swath of blue fabric hangs from the ceiling to the floor (the ocean), on which images and text are projected on at different moments, while centre stage is a triangle of fluorescent tube lights (the transatlantic slave trade route). A large piece of pink rock salt has a central position. There is a high wooden chair surrounded by green tropical plants in pots. The character of The Woman is wearing a long white cotton dress with black safety boots peeking out below.

In a section near the beginning of the piece entitled 'breaking the burden open', The Woman dons white safety gloves and her safety goggles, indicating that those audience members sitting close enough should do the same. With her sledgehammer she smashes the large rock of salt representing Europe (Fig. 2.3). She stops and collects herself before telling the audience that she has always lived in Europe, that 'Europe is awash in blood. Every penny of wealth, each brick of each intimidating building, the pavement slabs of quiet city streets and the soil beneath rolling green hillside is built on suffering, massacre, death. It is, and should be, a cursed continent' (Thompson 2019: 19). The ecology of Europe—the soil, the green hillside, the city streets—are shaped by colonial violence. White colonial patriarchy and capitalism are in The Woman's body as she breaks the salt, enacting violence and attempting to break the structures with the sledge-hammer.

Theresa J. May frames the body as the site of ecological experience: 'ecological issues, like the concerns central to feminism and postcolonial and multicultural theory, address injustices felt in the body—the body of

Fig. 2.3 Rochelle Rose in *Salt.* by Selina Thompson, Royal Court, London, 2019. (Photo: Johan Persson, courtesy of Selina Thompson Company)

experience, of community, of land' (2007: 101). The focus on the material and metaphorical body allows for theorisation of how identity and community shape injustices with ecological consequences. The body of The Woman holds the trauma and recounts the violence of the past. When she arrives in Ghana 'with the lens of the transatlantic slave trade over my eyes, pinning me to a mattress most days' (Thompson 2019: 35), she is there to grieve for the enslaved people that left there long ago, while her Nan's funeral is happening without her in Birmingham. She stores up her grief in her body and it paralyzes her. She waits for her period but it does not come. This is akin to what Emilia Quiñones-Otal calls the intersectional violence of colonialism that particularly affects women (2019: 678). In Mexico and Central America, artists are using their bodies as metaphors for this intersectional violence, where women's bodies are dominated territories within artworks. The Woman's body resists being dominated as she hammers the salt and smashes the structures that oppress her.

The Woman recounts instances of casual violence and racism she has experienced, jumping between years: 2011 being stared at in a cafe, then being called the n-word by a child in Bristol in 2014, repeating 'Europe pushes against me, I push back' (Thompson 21). The cargo ship Thompson boards from Belgium to Ghana is run by a White Italian captain ('the Master') who will not let Thompson and her filmmaker collaborator film any footage aboard the ship and makes it clear he does not agree with the project; 'had he known what our work was about, he would never have agreed to our presence on board' (24). They are the only women aboard with six White Italian officers and 19 Filipino crew members. In a section of the performance entitled 'The First Side: the Hold', Thompson recounts the experience of the three week journey in which they are not allowed on deck, stuck in windowless cabins, no sunshine, fresh air, phone or internet reception. The officers repeatedly use the n-word, while the 'Master' exhibits 'cartoon racism, impolite, brutish racism, not the smooth slick polite confused racism of my liberal friends. The shock it creates feels like a blow to the skull' (26–27). She grinds salt broken off from smashing Europe with a mortar and pestle as she tells of her experience on the ship, portraying the violence of the slave trade. 'On board, in our floating portion of Europe things are circling round and repeating themselves' (29). The oppressive structures of Europe are still pushing against her as she grinds down the salt.

As she crosses the Middle Passage, facing the hostile racism of officers she is forced to live with and dine with, the ghosts and residue of the violence pervades her mind. She is haunted:

> The trail of bleached bones stretching out across the middle passage, beneath my feet, added to every single day…
> At sea
> Passing over so much death
> Floating through it. (Thompson 2019: 29)

The ocean contains the social, political and the ecological consequences, memories and enforced forgetting of the slave trade. As commodities are continually shipped across oceans, as microplastics, pollution, oil spills and bodies of drowned refugees pervade the ecology of the sea, the violence of the slave trade is being reiterated in different formations. Sharpe articulates the way in which the traces of violence and death remain in the elements:

> But even if those Africans who were in the holds, who left something of their prior selves in those rooms as a trace to be discovered, and who passed through the doors of no return did not survive the holding and the sea, they, like us, are alive in hydrogen, in oxygen; in carbon, in phosphorous, and iron; in sodium and chlorine. This is what we know about those Africans thrown, jumped, dumped overboard in Middle Passage; they are with us still. (Sharpe 2016:19)

She invokes the materiality of elements in the air, land and sea as carrying the aliveness of murdered ancestors. This resonates with the multiple ecological meanings contained in the salt. It is memorialising the bodies lost in the ocean and the violence of their deaths. They are preserved in salt.

The semiotics of the enforced movements of The Woman on the cargo ship mirror the enforced movements of those enslaved, which continue, Sharpe argues, in contemporary formations: 'the forced movements of the migrant and the refugee, to the regulation of Black people in North American streets and neighborhoods, to those ongoing crossings of and drownings in the Mediterranean Sea, to the brutal colonial reimaginings of the slave ship and the ark; to the reappearances of the slave ship in everyday life in the form of the prison, the camp, and the school' (Sharpe 2016: 21). For Sharpe anti-Blackness is the total climate. Thompson draws these parallels in the sequence in which she tells of different instances of racism she has encountered before deciding to go on the journey. The shooting of the unarmed Michael Brown in Ferguson, Missouri, in 2014 was followed by more deaths, including Tamir Rice and Eric Garner: 'I am growing accustomed to a timeline, an endless feed of black pain, black rage and black people having to assert that black lives matter because black death is normal' (Thompson 2019: 20–21). In a total climate of anti-Blackness, the semiotics of the slave ship are repeating in different forms.

The police violence against Black bodies, the timeline full of Black death that Thompson experiences, is interconnected with environmental racism and injustices. In Beyoncé's popular *Formation* (2016) music video there is an image that strikingly connects these: Beyoncé sitting on a sinking police car amidst a flooded New Orleans, homes submerged in water all around her. Later she is standing triumphantly on the car, defiant. Near the end of the video, as she lays atop the car, her body pushes the car underwater. One of the last shots is her body sinking, water bubbling around her. She is exerting power and at the same time sacrificing herself, joining others in Black death. Like Thompson's eulogy at the end of *Salt.* when The Woman's body plunges into the abyss of the sea, 'Where no one should live. Where only death goes. I stay here, a floating grain of salt, a part of the grave' (2019: 50). I read Beyoncé atop the police car amidst a flooded New Orleans as a reference to the way the effects of climate change and ecological disaster (specifically Hurricane Katrina) are unevenly distributed along gender and racial lines, again repeating the semiotics of the slave ship. In the ecodramaturgy of the music video, the images are montaged with panning shots of New Orleans (flooded and not), graffiti that says 'stop shooting us', a line of police in riot gear facing off against a young Black boy dancing and images of Beyoncé and her dancers in formation referencing historic and contemporary images of Black culture from the American South. An intersectional ecological reading connects the history of racial segregation in the US, uneven urban development, a racist system of law enforcement and people who are politically marginalised based on gender, race and/or class to environmental disaster and the precarious weather events of climate change. As environmental scholar Caroline Finney contends, Hurricane Katrina highlighted 'the complex interaction of race and environment' (2014: 2) as certain neighbourhoods and communities were disproportionately affected. While Adeola and Picou point to a history of discriminatory housing practices, an 'enduring system of southern apartheid, involving racial segregation and consequent established patterns of community settlement of people of colour in less desirable, low-lying, flood-prone environments' meant therefore that communities of colour were hit hardest by the effects of the hurricane (Adeola and Picou 2017: 229). Ten years after the hurricane, 'recovery in African-American neighbourhoods hardest hit by flooding in Gentilly, the Lower Ninth Ward, and New Orleans East remains dismal relative to more affluent white neighbourhoods' (Adeola and Picou 230). May writes about the environmental justice issues exacerbated by the hurricane: 'as officials tallied the dead, my attention was arrested by the lock step between systemic social injustice and environmental lunacy. The so-called natural disaster has exposed the unnatural systems of domination buried under decades of business as usual, unearthing a white supremacist patriarchy through which racism, poverty, and environmental degradation are inseparably institutionalized' (2007: 127). This system of white supremacist patriarchy is what Thompson is pushing against and grieving the effects of. It is one that posits that social mobility, wealth and whiteness insulate against the effects of climate change.

The personal, autobiographical ecodramaturgy of *Salt.* uses the specific to, as Sharpe articulates, 'mourn and to illustrate the ways our individual lives are always swept up in the wake produced and determined, though not absolutely, by the afterlives of slavery' (Sharpe 2016: 8). *Salt.* is the personal story of Thompson's experiences, which she ties to the legacy of transatlantic slavery. To use the personal to excavate the social and historical formations of Black beings again counters what Saidiya Hartman calls the 'violence of abstraction' (Hartman 2008: 7). It pushes back against the constructions of gender, class and race that exclude Black women by tracing the oppressions of the colonial slave trade through her individual narrative. The abstraction of slavery and capitalist logics which dehumanise, racialise and genderise (Lugones 2012) is opposed through the naming and recounting of Thompson's feelings and experiences, biologically and represented in the lively ecology of salt. In this case, the personal is the ecological.

Arriving in Jamaica, The Woman sits on the high wood chair between two potted palms and tells us of the life and land that contains personal and political histories:

> Land of my mothers and fathers, and their mothers and fathers—except for one, who came from Montserrat. Land of my birth parents. Land of wood and water. Land of my blood. The island we left…
> It is
> Paradise.
> Fecundity defines Jamaica, it is everywhere, life, and I rest: fecundity, fertility, bursting and exploding tiny pale white green butterflies surrounding bright pink and orange flowers. (Thompson 43)

The ecology of Jamaica, the flora and fauna, is tied to the land of her ancestors, to slavery plantations, to suffering and inherited trauma. I read this as performing intersectional ecodramaturgy. She breaks the salt in an attempt to break the cycles and structures of oppression although she makes it clear she is not healed; this is still a burden that needs to be collectively carried. *Salt.* also performs Sharpe's wake work as it does 'not seek to explain or resolve the question of this [insistent Black] exclusion in terms of assimilation, inclusion, or civil or human rights, but rather depict aesthetically the impossibility of such resolutions by representing the paradoxes of blackness within and after the legacies of slavery's denial of Black humanity' (2016: 14). The salt on stage embodies some of these paradoxes, as 'salt was built on slavery' (Kurlansky 2002: 434). Salt as an industry was dependent on slave labour. Salt is also a commemoration of lives lost at sea; it is preserving the legacy of slavery in ecological terms. It evokes the journey across the sea that Thompson retraces. Audience members are invited to leave with a piece of salt: 'To take it is to make a commitment to live, a commitment to the radical space of not moving on, and all that it can open. Salt to heal, salt to remember, salt for your bath, for your nourishment, and above all for your wounds' (51–52). There is no easy conclusion and reconciliation, the paradoxes and pain are still present, preserved in the salt.

Salt. performs the relationship between colonialism and capitalism in its dramaturgy of retracing the slave trade's commodity chain. Lucy Tyler applies a global commodity chain analysis to *Salt.*, asserting that in its dramaturgy it reveals 'how racism is an ideology constructed by the slave trade's commodity chain. It allows Thompson to literally retrace the process in which value was "reflected" in subjects in a way that rendered humans into slaves and transformed slaves into labour power on arrival to the Americas' (Tyler 2019: 274). This ideology is echoed and reiterated in Thompson's contemporary experiences: the racist instances she details in the UK as well as her experience of the racist Master and crew aboard the cargo ship to Ghana who repeatedly use the n-word and even her experience travelling through airports where she was repeatedly searched and checked, pulled out of a queue in Hong Kong to be screened for disease five times. The Woman says: 'these experiences resonate with but pale in comparison to Bodies washing up from the Mediterranean Sea' (Thompson 2019: 41). The layering of time and the presence of the ongoing residue of slavery (and its afterlives) in the piece demonstrates that 'the history of capital is inextricable from the history of Atlantic chattel slavery' (Sharpe 2016: 5). The extractivist ideologies of capitalism link land dispossession and the dispossession of personhood in chattel slavery. Here, both land and 'slave' are categorized as inhuman matter, objectified and treated as commodities 'without subjective will or agency' (Yusoff 2018: 6). Yusoff further articulates the way in which the 'afterlives of slavery' (Hartman

1997) shape the earth through 'the extraction of gold, silver, salt, and copper to the massive transformation of ecologies in the movement of people, plants, and animals across territories, coupled with the intensive implantation of monocultures of indigo, sugar, tobacco, cotton, and other "alien" ecologies in the New World' (Yusoff 2018: 6). Salt can be added to this list as a commodity produced by slavery. By retracing the Transatlantic slave route, Thompson excavates these erased and ignored capitalist ecological afterlives of slavery.

Salt. and *Common Salt* address a history of forgetting the legacies and ongoing violence of capitalism and colonialism in ecological terms. *Common Salt* re-stories objects in the context of museums, such as the salt cellar surrounded by miniature British military figures. In *Salt.* Thompson's personal journey and embodied experience, typified by the smashing of the rock salt, is a way of tracing the past. The dramaturgy of making visible links between the ideologies of colonialism, capitalism, slavery and extraction contained in salt as a commodity is what I am calling intersectional ecologies. Here intersectionality moves beyond personal identity to the wider socio-political interlocking power structures that shape ecological relations.

Then She Said It

Then She Said It (2002) by Nigerian playwright Osonye Tess Onwueme illuminates the gendered effects of another global commodity: oil. Specifically, the play focuses on oil extraction in the Niger Delta and the intersections of environmental injustice and human rights injustice with women bearing a particularly violent burden. Onwueme is considered a second-generation playwright of the Nigerian National Theatre movement and is originally from Delta State in the Niger Delta region of Nigeria. Through weaving together the experiences of an array of characters, the play tells the story of women building solidarity to stand up to the intersectional oppressive forces which makes them, as Onwueme describes, 'the constant losers in the battle for oil and political dispensation of authority within the Nigerian context' (interviewed in Becker 2002: 28). An intersectional ecological analysis, through the character-driven, Brechtian-inspired dramaturgy, opens up the ways in which the effects of a colonial, extractivist ideology are gendered and raced. In researching the play, Onwueme spent time in the Niger Delta, speaking with the women there and learning about their lives and perspectives. She did this through a process of cultural sharing: finding intermediaries in the area who spoke the local language or dialect and then building relationships with the local market women. She would host evenings to bring the women together, with drinks and food, to share folktales, singing, dancing and experiences. Despite being from the region, Onwueme recognised the privilege she now has as a middle-class, educated woman holding a professorship at a university in the US. The class differences were important for her to acknowledge: 'I have come to realize in our own pursuit of women's empowerment as middle class and educated women… we tend to assume that we know it all. And we can position and speak for everyone else. And the unfortunate thing is that we know so little. Because the world of those women is so different from our own world. So we tend to interpret and read them from our own point of view, and I think that it is our own way of reproducing what we are trying to overhaul' (Onwueme interviewed in Becker 2002: 32). Her aim was to speak with, not talk for, the rural women and to bring their experiences to light as they are not often visible in national and international contexts. She also interviewed oil company executives and government officials to gain a fuller picture of the structures and actors involved.

The Niger Delta includes nine states with over 40 ethnic groups and a vast array of languages and dialects. They are generally what Rob Nixon refers to as 'micro-minorities' (2011), in that they have relatively small numbers compared to the overall population of Nigeria. The Niger Delta is the oil producing region of resource-cursed Nigeria and also one of the poorest areas of the country. Being resource-cursed is broadly a phenomenon in the Global South which applies to countries or regions rich in 'natural' resources that have value on the global market, and correlates to corruption, environmental injustice and social volatility. Nigeria depends heavily on oil and 'the petro-state has given rise, moreover, to a society in which 85 percent of the oil wealth goes to a mere 1 percent of the populace, almost none of whom belong to the micro-minorities who inhabit, ingest, and inhale the ecological devastation' (Nixon 2011: 106–107). The violence of sustaining such a petro-regime gained international attention when writer turned activist Ken Saro-Wiwa and the Ogoni Eight were executed on trumped-up murder charges for ostensibly interfering with the 'smooth economic running' of Shell's oil extraction. Their non-violent activism was in protest of the environmental devastation (and therefore economic and social devastation) in Ogoniland in the Niger Delta by Shell who was working in cooperation with Nigeria's military government.

Saro-Wiwa characterised the ecological violence as genocide of the Ogoni: 'The Ogoni country has been completely destroyed by the search for oil... All one sees and feels around is death. Environmental degradation has been a lethal weapon in the war against the indigenous Ogoni people' (Saro-Wiwa in Nixon 2011: 110–111). Their movement was successful as it drove Shell out of the area, but it cost them their lives (and the 2000 Ogoni who were murdered and villages burned in military raids). Shell relocated to other parts of the Niger Delta, reiterating the same practices while investing in greenwashing in Europe to help its reputation (Nixon 2011).

While Saro-Wiwa's story tends to dominate the Global North's cultural imaginary of the Niger Delta, Onwueme's play is about rural women who have been 'completely marginalized and terrorized in this whole process. And they too have become quite militant in their own resistance' (Onwueme in Becker 2002: 29). In *Then She Said It*, Onwueme imagines a fictional nation that stands in for Nigeria (Hungeria) and divides the stage between the Oil Club (with a sign saying 'Restricted Area. Keep Off!'), and the Market Square, which is the hub of community life for women. Here the tensions between the communities bearing the brunt of the ecological consequences of oil extraction and those gaining the economic wealth are played out through a character-centric dramaturgy. As Onwueme notes 'the characters embody ideas or are symbols of ideas that collide with one another, conflict with one another. And through that kind of agency, one is able to explore multiple dimensions of meaning' (Onwueme interviewed in Becker 2002: 36). The younger characters multi-role, embodying a number of different perspectives that overlap and change. In a note at the beginning of the text, Onwueme writes that all the characters are named after bodies of water or rivers, 'like the swift currents of water that defy limiting boundaries, these promethean characters both embody and project the dynamic change or movement that transforms the life and consciousness of a people, hanging on the cliff of existence and power' (2002: ix). The characters represent a vast array of different people involved and affected by the extraction and exploitation.

The character of Atlantic is described as a white foreign oil company director, who stays behind the guarded gates of the Oil Club where he meets with the Governmental Official and Traditional Chief to strike up deals that protect their own interests and maintain control of the oil fields. Near the beginning of the play he makes a toast: 'life in Africa is cheap', referring to those 'expendable' populations as the cheap cost of doing business. Atlantic represents the foreign oil companies (such as Shell—called Shame in the play—Chevron, Mobil and Texaco) and the transatlantic ecological relationships imbued in oil consumption. Italo Calvino evokes these relationships as he fills his gas tank in 'The Petrol Pump':

> as I fill my tank at the self-service station a bubble of gas swells up in a black lake buried beneath the Persian Gulf, an emir silently raises hands hidden in wide white sleeves and folds them on his chest, in a skyscraper an Exxon computer is crunching numbers, far out to sea a cargo fleet gets the order to change course. (Calvino 1995: 173–174)

Instead of a black lake buried beneath the Persian Gulf, it is a black lake in the Niger Delta, with Shell and other oil companies building pipelines that leak, polluting the soil, water and air, devastating the agriculture in the area (Amnesty International et al. 2020). Nixon asserts the treatment of the Niger Delta by Shell is based in racism: 'the company waives onshore drilling standards that it routinely upholds elsewhere. Indeed, 40 percent of all Shell oil spills worldwide have occurred in Nigeria.' (Nixon 2011: 113). Shell does not pay local authorities in the Niger Delta the lucrative rent it pays locally in its operations in the Northern hemisphere. Shell flared natural gas in its production at alarmingly high rates compared to the UK and the US: 76% of the natural gas from petroleum production in Nigeria, compared to 4.3% in the UK and 0.6% in the US in 1995, which as a 'toxic practice foreshortened the life expectancy of the delta peoples' (Nixon 2011: 113). The effects of this environmental racism are not isolated to the Niger Delta as the carbon and methane emissions from the flaring of the oil fields is a significant contributor to climate change.

In the play, Atlantic treats the women with the same exploitative ideology as the land, water and air. He asserts a sense of ownership, using them for this his own benefit and discarding them when he is finished. The sexual violence, including rape and human trafficking, is interconnected with the ecological violence. The character Oshun, a young woman who has been a sex worker to Atlantic who then refuses to pay her, stands up to him saying: 'all you want is just to kick us around and out…to clean up your mess…To drain…To drill… No, man. Not any more! You pour it? You pay for it! You pump it? You pay for it. You mess it up? *You* clean it up' (Onwueme 2002: 52). The language of the sex acts and the oil extraction are one and the same here. Exploitation and objectification happen to both the women and the environment, while Atlantic repeats that he is only protecting his business interests, that his company has a contract with the government who are to blame for the lack of fuel, jobs and livelihoods for the local 'savages'. All the while, he is working closely with the Government Official and the Traditional Chief to enhance their personal wealth and oil interests because both the ecology of the area and the rural women are expendable for profit.

Atlantic's perspective calls for an understanding of how power formations shape everyday lives within the context of ongoing and historical ecocide as well as contemporary climate crisis. Intersectional ecological thinking can account for complex and overlapping oppressions and power, 'furthering our understanding of how a person's relationship with the environment (in the Global South or North) is not completely dependent on any one aspect of their lives, whether gender, race, class, sexuality or age but rather a combination of all of the above and more besides' (Kings 2017: 71). Without essentialising, Cherríe Moraga and Gloria Anzaldúa write in *The Bridge Called My Back* (1981) that 'the physical realities of our lives—our skin colour, the land or concrete we grew up on, our sexual longing all fuse to create a politic born out of necessity' (23). These material conditions and physical realities mean the rural women in the Niger Delta cannot separate their identities into distinctive colonial-structured categories, rather their intermeshed lived experience gives rise to an ecofeminist politics and activism, connecting land and place to labour, race, sexuality, gender and immigration status—all informing each other.

There are two generations of women who band together to stand up for their rights. There is a generation of 'mothers' led by Niger and Benue, both widows in their forties who lead the market women. They lost husbands and sons in the violent conflicts with the police and military who protect the oil extraction operations. The younger generation of 'daughters' is led by Obida and Oshun, representing the youth of the area who are educated but unemployed. Both are survivors of sexual violence. While waiting in line for fuel (of which there is a shortage) these two generations come together over their shared ecological marginalisation:

Obida: They're not even killing us alone. The trees too!
Niger: Our farmlands!
Benue: And rivers!
Niger: The environment.
Benue: Polluting.
Niger: Polluting the land, the rivers, our entire environment…

Women:	All polluted.
Obida:	You said it.
Niger:	Then she said it.
Obida:	They've killed everything with their oil pollution and spillage. We cannot breathe clean air. Fishes die or get fried in the polluted simmering rivers. Water—water everywhere. But we have no clean water to drink! And now we lose the land too?
Niger:	No firewood because the plants and trees are soaked in oil. What do they expect us to cook with? (Onwueme 2002: 15).

Onwueme uses choral work to develop the solidarity between women to build a movement and represent the community at large. In the scene above they are relating their shared experience of social and environmental injustices. This happens again as they wait in line to apply for one job (for a guard) that is posted outside the Oil Club. Many of them have degrees but there is no employment and the few jobs that are available in relation to the oil extraction do not often go to those in the local community. The chorus becomes more vocal as the women are galvanized as activists, representing the community strength and solidarity. It was Onwueme's intention to involve community members as the chorus when staging the production. It was important to her to bring the play to the people who inspired it. It first premiered at the State Cultural Centre, Calabar, Cross-River State, Nigeria, and was then selected for Command Performances by the Delta State Government.

The women in the play eventually decide the sexual violence, loss of agriculture, lack of jobs, poverty, fuel shortages, pollution, displacement, corruption and loss of their husbands and sons is too much to bear. They militarise and decide to control their own resources; they kidnap Atlantic for ransom and set fire to an oil pipeline. They are arrested and tried but manage to overtake the courtroom and defend their actions, confronting the characters representing the interlocking structures of oppression they have endured: environmental racism of the oil companies who treat the people and land as disposable; the government for the lack of infrastructure in their area including fuel shortages, lack of schools and employment opportunities, for hoarding the 'compensation' paid by the oil companies and inflicting violence on them through military means; and the Traditional Chief who sells out his people to further his own interest. These characters represent mechanisms of global capitalism which devastate the Niger Delta and disproportionately burden the rural women. These mechanisms are seen in the play through unethical business deals, unbridled pollution of the environment for profit, racism and sexual violence perpetrated by the powerful men (Traditional Chief, Government Official and Atlantic) who can act without consequence because the structures of power were made for them. Enacting and representing these structures through different characters Onwueme elucidates an intersectional understanding of the multitude of effects of oil extraction in the Niger Delta.

The plays and performances in this chapter uncover and illuminate some of the interlocking forces that shape social-political-ecological relations and the way in which they are intermeshed and inseparable. The normalisation of these relations—from colonial land development to salt taxation and production through slavery to filling up a car with petrol—hides the biopolitical and indexical connections. Through tracing the injustices, artistic and cultural representations can reimagine and reframe these relationships as in the Green Belt Movement, *Common Salt*, *Salt.* and *Then She Said It.*

Intersectionality unravels the entanglements around how power is produced, consolidated and enacted, paying particular attention to how oppressions might overlap and accumulate. Intersectional ecologies must therefore pay attention to how power and privilege 'whitewash' ecological ideas and affect our ability to understand experiences outside our own. As I write this, a plastic straw debate is happening in the UK, cities in the US and many other places, which disability activists contend is another way that mainstream environmentalism is excluding them and silencing their voices. The debate exposes the myriad of fusions that everyday plastic straws contain. Flexible plastic straws were originally designed as a medical aid and for those that rely on them for eating and drinking; a suitable 'green' alternative that is as flexible, strong and allows for biting has not yet been developed. UK performance maker Jess Thom (Tourettes Hero) has written about the way in which the debate and reduction in access to plastic straws sets up a false binary between individual needs versus 'collective good', again erasing disabled people from the narrative. 'Straws have become a focus of the anti-plastic campaign because to many people they seem like a frivolous luxury. But our oceans are full of other damaging plastics that most people would see as essential, like medical gloves for example' (Thom 2018). What the campaign seems to be missing is that for some people, plastic straws are an essential medical tool. Yet this perspective is not widely represented in memes, social media call outs and campaigns or nature documentaries on the subject that seem to have galvanised people around this issue. All-out plastic straw bans, before a suitable alternative has been developed, would hinder the independence, access and ability to fully participate in social activities for many people with disabilities, while not getting to the heart of the problem about plastic waste. A ban, without a way of ensuring the full participation of the disabled community, would be short-sighted and continue the normative (and exclusionary) logic of many other environmental 'solutions'. These 'solutions' are premised on a particular idea of the 'human' with a normative body, non-disabled and socially mobile (for whom the use of a straw is a luxury rather than a necessity that allows them to participate in social activities such as dining out for example). It also presumes a relationship to ecology based on consumption and disposal, without attending to how waste streams are treated (including the exporting of waste to less 'developed' countries). Mindful of avoiding anti-human ecological thinking, intersectional ecologies are a critical tool for unpacking how we can extend an understanding of marginalisation, asking the other questions about who or what is being erased or excluded in this story.

Intersectional ecologies are a way of thinking about and analysing eco-dramaturgies, which I continue through the rest of this book. In Chap. 3, I continue thinking about feminist theory, discussing how ecomaterialism offers a way of undoing the reductive hierarchy of heteropatriarchal anthropocentrism through performing vibrancy (or bioperformativity). I expand on some of the thinking about the liveliness of matter and materiality started in this chapter, inspired and guided by performances from Joseph Beuys, Fevered Sleep and Lucy + Jorge Orta.

Notes

1. The British East India Company enforced the salt tax in India from the eighteenth century. They established customs checkpoints to stop the smuggling of salt. These checkpoints were then connected and expanded to form a Customs Line made up of a 'fourteen-foot-high, twelve-foot-thick thorn hedge on the western side of Bengal to prevent the entry of contraband salt' (Kurlansky 2002: 339). When the British Crown took over 'control' of India in 1858, the impenetrable Customs Line was expanded to 2500 miles across India, employing 1200 people by 1870 (Kurlansky 2002: 339–340). India 'gained independence' in 1947 and the salt tax was abolished in 1950.
2. Examples include *The Contingency Plan* (Waters 2009), *Earthquakes in London* (Bartlett 2010), *Greenland* (Buffini, Skinner, Charman, Thorne

2011), *The Heretic* (Bean 2011), *Ten Billion* (Emmott 2012), *2071* (Rapley and Macmillan 2014), *Pastoral* (Eccleshare 2013), *Lungs* (Macmillan 2014), *F*ck the Polar Bears* (Ronder 2015) and *Oil* (Hickson 2016).
3. Statistics in the play from WECAN: https://www.wecaninternational.org/why-women
4. The Windrush Generation is around 500,000 people who migrated to the UK from Caribbean colonies between 1948 and 1971. Many of these people were children travelling with their parents, without their own passports or paperwork. Changes to immigration law in 2012 created a 'hostile environment' and required people to provide documentation of their right to work and access services. In 2010, the UK Home Office destroyed the Windrush landing cards leaving many people unable to 'prove' their right to remain in the UK (Vargha 2018). Detentions and deportations ensued. An enquiry in the matter found 'profound institutional failure' based on racial discrimination and financial compensation was offered to those affected (BBC News 2020).

References

Adeola, Francis O., and J. Steven Picou. 2017. Hurricane Katrina-Linked Environmental Injustice: Race, Class, and Place Differentials in Attitudes. *Disasters* 41 (2): 228–257.

Amnesty International, ERA/FoEN, Friends of the Earth Europe, and Milieudefensie. 2020. Nigeria: No Clean-Up, No Justice: An Evaluation of the Implementation of UNEP's Environmental Assessment of Ogoniland, Nine Years On. https://www.amnesty.org/download/Documents/AFR4425142020ENGLISH.PDF.

BBC News. 2020. Windrush Scandal: Home Office Showed 'ignorance' of Race. *BBC News*, March 19. https://www.bbc.co.uk/news/uk-politics-51961933?intlink_from_url=https://www.bbc.co.uk/news/topics/c9vwmzw7n7lt/windrush-scandal&link_location=live-reporting-story.

Becker, Becky, and Osonye Tess Onwueme. 2002. Then, She Said It!: An Interview with Playwright, Tess Onwueme. *Feminist Teacher* 14 (1): 27–40.

Bilodeau, Chantal. 2019. *It Starts With Me*. Climate Change Theatre Action.

Calvino, Italo. 1995. The Petrol Pump. In *Numbers in the Dark: And Other Stories*, 170–175. Translated by Tim Parks. New York: Pantheon Books.

Collins, Patricia Hill. 2000. *Black Feminist Thought: Knowledge, Consciousness, and the Politics of Empowerment*. Revised 10th anniversary ed. New York: Routledge.

Crenshaw, Kimberle. 1989. Demarginalizing the Intersection of Race and Sex: A Black Feminist Critique of Antidiscrimination Doctrine, Feminist Theory and Antiracist Politics. *The University of Chicago Legal Forum* 140: 139–167.

Dater, Alan, and Lisa Merton. 2008. *Taking Root*. DVD. Directed by Alan Dater and Lisa Merton. Newburgh, NY: New Day Films.

Davis, Kathy. 2008. Intersectionality as Buzzword: A Sociology of Science Perspective on What Makes a Feminist Theory Successful. *Feminist Theory*. 9 (1): 67–86.

Finney, Carolyn. 2014. *Black Faces, White Spaces: Reimagining the Relationship of African Americans to the Great Outdoors*. Chapel Hill: The University of North Carolina Press.

Ghelani, Sheila, and Sue Palmer. 2018. *Common Salt*. Performance. London: Battersea Arts Centre, June 7–9.

———. 2020a. *Common Salt*. Performance. Reading, UK: Museum of English Rural Life, University of Reading, January 20.

———. 2020b. *Common Salt*. Flyer. University of Reading, January 20.

Ghelani, Sheila, and Lisa Woynarski. 2018. Ecology, Colonialism and Agency: Objects in Dialogue. Presentation Creative Climate Symposium, *Birkbeck*, London, May 8.

Green Belt Movement. 2016. The Green Belt Movement Annual Report. The Green Belt Movement. Accessed February 21, 2020. http://www.greenbeltmovement.org/sites/greenbeltmovement.org/files/compressed_488128702.pdf.

Green, Cathy, Susan Joekes, and Melissa Leach. 1998. Questionable Links: Approaches to Gender in Environmental Research and Policy. In *Feminist Visions of Development: Gender, Analysis and Policy*, ed. Cecile Jackson and Ruth Pearson, 259–283. London; New York: Routledge.

Hancock, Ange-Marie. 2016. *Intersectionality: An Intellectual History*. New York; Oxford: Oxford University Press.

Hartman, Saidiya V. 1997. *Scenes of Subjection: Terror, Slavery, and Self-Making in Nineteenth-Century America*. New York; Oxford: Oxford University Press.

———. 2008. *Lose Your Mother: A Journey along the Atlantic Slave Route*. New York: Farrar, Straus and Giroux.

Kaijser, Anna, and Annica Kronsell. 2014. Climate Change through the Lens of Intersectionality. *Environmental Politics* 23 (3): 417–433.

Karanja, MaryAnn. 2019. *Birthday Suit*. Climate Change Theatre Action.

Kings, A.E. 2017. Intersectionality and the Changing Face of Ecofeminism. *Ethics & the Environment* 22 (1): 63–87.

Kurlansky, Mark. 2002. *Salt: A World History*. London: Vintage.

Luft, R.E., and J. Ward. 2009. Toward an Intersectionality Just out of Reach: Confronting Challenges to Intersectional Practice. *Advances in Gender Research* 13: 9–37.

Lugones, Maria. 2012. Methodological Notes toward a Decolonial Feminism. In *Decolonizing Epistemologies: Latina/o Theology and Philosophy*, Transdisciplinary Theological Colloquia, ed. Ada María Isasi-Díaz and Eduardo Mendieta, 1st ed., 68–86. New York: Fordham University Press.

———. 2014. Radical Multiculturalism and Women of Color Feminisms. *JCRT* 13 (1): 68–80.

Matsuda, Mari. 1991. Beside My Sister, Facing the Enemy: Legal Theory out of Coalition. *Stanford Law Review* 43 (6): 1183.

May, Theresa J. 2007. Beyond Bambi: Toward a Dangerous Ecocriticism in Theatre Studies. *Theatre Topics* 17 (2): 95–110.

Mies, Maria, and Vandana Shiva. 1993. *Ecofeminism*. New Delhi: London: Kali for Women; Zed Books.

Moraga, Cherríe, and Gloria Anzaldúa, eds. 1981. *This Bridge Called My Back: Writings by Radical Women of Color*. 1st ed. Watertown, MA.: Persephone Press.

Moxham, Roy. 2002. *The Great Hedge of India*. London: Constable.

Museum of English Rural Life. 2020. *Common Salt Pop Up Exhibition*. Exhibition. Reading, UK: Museum of English Rural Life, University of Reading, January 20.

Nixon, Rob. 2011. *Slow Violence and the Environmentalism of the Poor*. Cambridge: Harvard University Press.

Onwueme, Osonye Tess. 2002. *Then She Said It*. San Francisco: African Heritage Press.

Quiñones-Otal, Emilia. 2019. Women's Bodies as Dominated Territories: Intersectionality and Performance in Contemporary Art from Mexico, Central America and the Hispanic Caribbean. *Arte, Individuo y Sociedad* 31 (3): 677–693.

Sharpe, Christina. 2016. *In the Wake: On Blackness and Being*. Durham: Duke University Press.

Singh, Amit. 2016. Poll Shows Brits Are Proud of Colonialism—Clearly They Haven't Heard of These Colonial Crimes. *The Independent*, January 20. http://www.independent.co.uk/voices/poll-shows-brits-are-proud-of-colonialism-clearly-they-havent-heard-of-these-colonial-crimes-a6823151.html.

Thom, Jess. 2018. #INeedStraws. *Touretteshero*, July 11. Accessed August 06, 2018. https://www.touretteshero.com/2018/07/11/ineedstraws/.

Thompson, Selina. 2019. *Salt*. London: Faber & Faber.

Tyler, Lucy. 2019. Work in Progress: English Play Development Under Neoliberalism, 2000–2019. PhD diss., London: Royal Central School of Speech & Drama.

Vargha, Dora. 2018. Windrush Scandal: A Historian on Why Destroying Archives Is Never a Good Idea. *The Conversation*, April 24. https://theconversation.com/windrush-scandal-a-historian-on-why-destroying-archives-is-never-a-good-idea-95481.

Velez, Emma D. 2019. Decolonial Feminism at the Intersection: A Critical Reflection on the Relationship Between Decolonial Feminism and Intersectionality. *The Journal of Speculative Philosophy* 33 (3): 390–406.

Yusoff, Kathryn. 2018. *A Billion Black Anthropocenes or None*. Minneapolis: University of Minnesota Press.

CHAPTER 3

More-than-Human Matters: Bioperformativity

Trees, a polar bear costume, ice, concrete, felt, skin, fur and weather. These are things in the performances discussed in this chapter *(7000 Oaks, It's the Skin You're Living In, Symphony for Absent Wildlife* and *The Weather Factory*) that have the potential to create effects on the world at large and in relation to humans. They can question the underpinning logics of the binary divisions between human/nonhuman and nature/culture. They are active things that have what Jane Bennet calls 'thing-power: the curious ability of inanimate things to animate, to act, to produce effects dramatic and subtle' (Bennett 2010: 6). In other words, they have the capacity for agency. One way of thinking about ecodramaturgies is recognising how more-than-human matter performs, such as the things in the performances of this chapter. Performativity[1] is an active word, as it creates action and effects, broadly construed solely in relation to the human. This anthropocentrism is limiting and reductive, as Karen Barad suggests: 'performativity has been figured (almost exclusively) as a human affair; humans are its subject matter, its sole matter of concern' (Barad 2011: 122). What I call bioperformativity counters this anthropocentrism by understanding the biological/material and the performative effects of things. Not all of these things have the same agency or operate in the same way. Some of them are more complex and layered as their agency is distributed across assemblages of human and more-than-human 'things' as Bennett notes (2010). The aim here is not to suggest that these things perform in the

L. Woynarski, *Ecodramaturgies*, New Dramaturgies,
https://doi.org/10.1007/978-3-030-55853-6_3

same way but to acknowledge their performativity and capacity to create effects, in different contexts, in different degrees and kinds.

Performativity creates awareness of, as Elin Diamond writes, 'performance itself as a contested space, where meanings and desires are generated, occluded, and of course multiply interpreted' (Diamond 1996: 4). This is in conjunction with the concerns of this book: understanding how meaning is generated and interpreted in theatre and performance in ecological terms. Diamond configures performativity as a means of interpretation and critique, 'as soon as performativity comes to rest on a performance, questions of embodiment, of social relations, or ideological interpellations, of emotional and political effects, all become discussable' (Diamond 1996: 4). For ecodramaturgies, configuring more-than-human agency as performative allows for the problematising of categories and discussion of (ecological) meaning generated in performance. Bioperformativity is intended to critique and interpret the way the human and more-than-human are categorised and constituted in performance, acknowledging more-than-human performances through drawing attention to embodied ecological relationships, emotions, ideologies and political effects.

Trees as Performers

The thing-power of trees in tree planting projects can act as what Rob Nixon refers to as 'a symbolic hub for political resistance' (Nixon 2011: 132) as discussed in the Green Belt Movement in Chap. 2. As a symbol and material, trees exercise both material and narrative agency. They demonstrate the way ecodramaturgies can contribute to the process of excavating the intricate interrelationships between humans and more-than-humans, and in turn, how this thought process can lead to more ecologically equitable modes of being. Through an ecomaterialism framework, trees and other things can create powerful narratives as matter, images and metaphors for ecological thinking. This is an ecodramaturgical strategy for revealing the performativity of the more-than-human and non-anthropocentric ways of thinking about theatre and performance.

Trees are performative and have agency. Trees are pervasive symbols in ecological theatre and performance, and ecological imagery more broadly. The material agency of trees and their performative framing in artworks enact *bioperformativity*, making them enduring metaphors, symbols and actants in ecological visual language and narratives. Take, for example, Joseph Beuys' *7000 Oaks* (1982),[2] which is arguably one of the most famous artworks involving the planting of trees as performance. For the art festival Documenta 7 in Kassel, Germany, in 1982, Joseph Beuys planted the first of 7000 oak trees. The piece was intended to raise ecological awareness and renew the urban environment. Each tree was paired with a basalt stone, 7000 of which were piled up outside Museum Fridericianum to represent the trees to be planted. The piece was completed after his death in 1986 and has since spread to other cities. Beuys conceived of *7000 Oaks* as a regenerative project, aimed at restoring the biosphere by planting more trees in urban areas: 'I believe that planting these oaks is necessary not only in biospheric terms, that is to say, in the context of matter and ecology, but in that it will raise ecological consciousness' (in Beuys et al. 1982). The oak tree was also symbolic of his work of 'regenerating the life of humankind' (Beuys 2012: 167). For him, oaks trees have 'always been a form of sculpture, a symbol for this planet ever since the Druids, who are called after the oak' (Beuys 2012: 167). As Beuys points out, the oak tree has both metaphorical and material agency as it produces effects in biospheric terms: it converts CO_2 to oxygen, provides a nesting space for birds and food for insects, absorbs nutrients from the soil and slowly cracks the surrounding concrete. In these ways, trees act in a human-nonhuman assemblage, creating subtle reverberations and repercussions, both material and immaterial (such as ecological consciousness raising). Beuys locates the art and aesthetic experience in biospheric and social terms. By situating the tree as art, he draws attention to the way the tree performs and exercises material agency. To think of Beuys' work as enacting bioperformativity is necessary to question and critique the idea of the human and resist anthropocentrism, recognising the material agency of the more-than-human actants involved.

My neologism *bioperformativity* is about identifying non-anthropocentric ecodramaturgies through the recognition of the agency of the more-than-human, or the way the more-than-human performs as material and representation. Agency is not employed here as 'intention', 'choice' or audience agency as it has been positioned within immersive and participatory theatre scholarship.[3] Rather, it is configured as the ability to produce effects and to affect, influence or to make a difference to something, not necessarily within the realm of the human. 'So agency is not something possessed by humans, or non-humans for that matter. It is an enactment', asserts philosopher Karen Barad, 'and it enlists, if you will, "non-humans" as well as "humans"' (interviewed in Dolphijn and Tuin 2012: 55). In *7000 Oaks*, the trees enact bioperformativity because they create physical effects and symbolic effects as part of high-profile artwork aimed at raising 'ecological consciousness'.

The performance of Beuys' oak trees is a way to consider our understanding of the agency of a tree and the compulsion to plant them within environmental movements. The ecological agency of trees is evident through their use as a recurrent symbol and popular image within ecological discourses, ecocritical writings and ecological performance works. Historian Richard Hayman contends that this fascination with trees is because of their metaphoric potential and material presence, which is paradoxically contrasted by our behaviour towards them: 'no living things have had more impact on human sensibility than trees. Trees are special. They are bigger than us both physically and metaphorically, but we couple our reverence for them with a relentless destruction of forests' (2003: 1). Not only do humans affect and exercise agency towards trees, trees influence and give shape to humans. *7000 Oaks* acknowledges trees as active matter in the biosphere and on the human, enacting an ecological sensibility. The oak tree can be considered an artist similar to rivers as Jeffery Cohen describes:

> The Mississippi is an earth artist, but its projects take so long to execute that humans have a difficult time discerning their genius. The river composes with ice, stone, potent flows of water, heterogeneous biosystems, and tumbling sediment...An incessant flow of objects, animals, elements, and forces not reducible to human use-value, the powerful river exerts a relentless agency easily readable in its engendered worlds. (Cohen 2013: xix)

Trees also have agency in their recurrence as an 'ecological' symbol, sometimes problematically standing in for romantic ideas of nature, leading towards the fetishisation of tree planting within environmentalism.[4]

At the same time trees are more than romanticised images of ecology, they can be site-specific performers, cartographic records, living tissues of time, and points of reference in place and hybrid time species (Heddon 2013). Deirdre Heddon specifically addresses trees as both symbol and material in *Forest Pitch* (2012) by Craig Coulthard. The piece involved football matches with teams made up of people newly migrated to Scotland. The pitch was cleared in the middle of a forest on the Scottish Borders, a forest of commercially grown monoculture spruce trees. After the matches, native tree species were planted along the white lines of the pitch. Heddon draws out some of the inherent tensions and paradoxes in the piece such as the planting of native tree species while welcoming 'new

Scots' and cutting down trees to plant fewer trees (2016). Ideas of nativism, nationalism and the compulsion or fetishisation of tree planting are all running through the work. Heddon also considers the World Wildlife Fund (WWF) campaign 'Plant a tree seed, save a planet' as an example of this compulsion to plant trees. The campaign instructs people to plant a seed in three easy steps: bend down, make a hole and then cover the hole, although they do also ask that people choose a spot with light and space, and a native species of tree (2014). The campaign then goes on to suggest that the planting of a seed is not only a grand gesture towards saving the planet, but it also plants an idea: small-scale actions can fruit large-scale effects. Unlike the Green Belt Movement in Kenya which demonstrates the tree planting can produce effects through engagement with specific communities and locations and align with attendant socio-political movements, the WWF campaign presents a one-size fits all solution. The reductive simplification of the instructions of the WWF seems to suggest that the simple act of planting a tree is key to 'solving' the global ecological crisis without attending to the structures and logics that produced it. Trees take on a mythic power in this campaign: a single tree is able to undo immeasurable environmental damage. The inference seems to be that planting a tree will provide an alibi, as it will make up for all the other ecologically harmful effects of ideologies and power relations in the Global North (where the campaign seemed to be aimed). Here, trees are not only performing in a biophysical sense, they also perform within the construction of environmental responsibility in the public realm. Trees perform assumptions of 'doing good' where planting a tree becomes a selfless gesture with a big impact, such as in the WWF campaign. While the symbolic value of trees has been misused by campaigns like the WWF to absolve those in the Global North of ecological responsibility, many artists have used the performance of planting of trees (and other plants) as affective images that can problematise human and more-than-human ecological relationships as detailed below.

Within the recent history of 'eco art', artists have framed the bioprocesses of trees and plants as performance. More than just large-scale earthworks or 'environmental art', these works aim to 'restore' or 'remediate' a site and/or ecosystem. Sue Spaid describes this kind of work as 'ecovention': 'Coined in 1999, the term ecovention (ecology + invention) describes an artist-initiated project that employs an inventive strategy to physically transform a local ecology' (2002: 1). For example, in 1978 artist Alan Sonfist planted a native forest in the heart of Greenwich Village,

New York, in an effort to draw attention to the way in which humans can restore a landscape. He called the piece *Time Landscape*. Then in 1982, Agnes Denes made *Wheatfield—A Confrontation,* also in New York, in which she grew and harvested wheat on a site of rubble from the construction of the World Trade Centre in Manhattan. In 1991, Mel Chin worked with scientists to plant a large target-shaped garden on the site of the toxic waste in Minnesota in *Revival Field*. They planted hyperaccumulator plants, which remove toxicity from soil, helping to remediate the site. In 1992, Kathryn Miller performed *Seed Bombing the Landscape,* in which she used compressed soil and native seeds moulded into bomb shapes and flung them into degraded landscapes in need of vegetation. The idea was that the bombs would release the seeds into the landscape and an abundance of native flora would flourish. Artist Meghan Moe Beitiks views plants as co-performers, active participants in her work (2018). Her piece *The Plant is Present* (2011) involved people sitting in a chair opposite a potted *sansevieria trifasciata* (snake plant) for as long as they liked, inspired by Marina Abramović's *The Artist is Present* (2010). Writing about nonhuman performances in human-framed artworks, she refers to the way the plants perform: 'non-human entities, having presence, perform alongside human entities, through their own historicity and life cycles' (2013: 7). In all of these works, trees and plants perform in human-nonhuman confederations that affect and shape each other. Their performance is constituted by their capacity for agency. These works also comment on urban/nature binaries, implicitly critiquing them by using the more-than-human performance as a way of examining contested sites and distinctions. The artworks frame the more-than-human performers as actants in the multifarious ecological relationships of each site. The sites demonstrate the performance of the human in changing the ecological make-up of a place, and the performance of the human-nonhuman assemblages that are destroying, remediating or revitalising them. These assemblages are enacting material-ecological agency, or bioperformativity.

Ecodramaturgies as Non-Anthropocentric

Fevered Sleep's *It's the Skin You're Living In* (2013), Lucy + Jorge Orta's *Symphony for Absent Wildlife* (2014) and Fevered Sleep's *The Weather Factory* (2010), each extends the concept of bioperformativity by expanding and questioning the idea of what it means to be human. By focusing on the power relations and socio-political forces that shape our sense of the human and our relationship to the weather, climate and land, ecodramaturgies can displace a reductive idea of anthropocentrism. Each of these three performances features lively matter and material that enact performative agency through costumes, editing, scenography and sound.

Theorising ecodramaturgies as non-anthropocentric offers an opportunity to productively critique the perceived separation between humans and the more-than-human world, providing a new way of critically thinking about theatre and performance. This, of course, is a performative self-contradiction: a human elaborating a non-anthropocentric performance theory, as Jane Bennett (2010) identifies. Baz Kershaw also cautions that writing about theatre and performance ecology is performative as it 'could be reproducing the very pathology it wants to question: the exploitation and degradation of the Earth's environment by humankind' (2007: 300). However, this performative contradiction is a generative one as it can call attention to the way in which we as humans are inextricably embedded in the more-than-human world. Bioperformativity draws on ecomaterialist theories to revise 'operative notions of matter, life, self, self-interest, will, and agency' (Bennett 2010: ix). By opening up ecodramaturgies beyond the human, I aim for expansiveness, messiness and complexity in performance and theatre work that acknowledges the capacity for agency in the more-than-human without inscribing the nature/culture binary or configuring humans at the top of a strict, reductive hierarchy of being within a mechanistic worldview. For Una Chaudhuri, a 'new ecological consciousness' will be made by breaking 'the fifth wall of anthropocentrism' developing an 'atmospheric consciousness' in theatre that looks beyond the social world of the human, up to the ecological world (Chaudhuri 2016: n.p.). From this position, the discussion in the remainder of this chapter is in four sections: narrative agency in *It's the Skin You're Living In;* the queer ecological revisioning of binaries; the non-anthropocentric ecodramaturgies in *Symphony for Absent Wildlife*; and weathering and expanding the human/climate relationship in *The Weather Factory.*

It's the Skin You're Living In

It's the Skin You're Living In, a short film by Fevered Sleep, begins with a series of captions:

> You are watching a film_
> that concerns itself with_
> the matter of connectedness_
> with ice and flesh_

mountains and concrete_
migration and walking_
kinship and otherness_
here and there_
present and future_
fur and skin_ (Harradine 2013)

These captions set up the idea that the film is not about binaries but about the interconnections between seemingly distinct things. The captions also allude to the structure of the film, which is a series of short shots, sometimes in split screen, with no dialogue. The first shot is a wide shot of snowy mountains in split screen with the sound of wind and snow blowing across a frozen Arctic landscape. The right side of the screen cuts to a medium shot and we see a white polar bear on all fours, alone in the snowy landscape. The images reverse and the polar bear slowly moves. The film cuts to a closer shot of the polar bear standing on two legs and its movements slowly reveal it to be a human wearing a fairly realistic polar bear costume. Although it is difficult to pinpoint the 'human' moment, there is a quality in the slow movement that makes it clear this is a human performing as a polar bear. The ambiguity opens up space to consider the relationship between human, animal and climate. The human performance of a polar bear becomes clearer in a tight shot of the bear as it begins to walk on two legs towards the camera, the seams of the costume evident. With his back to the camera the bear figure begins to flap his arms, dislodging the arm coverings of the costume. The white male-presenting figure under the costume becomes evident. Another cut reveals the bear figure from a different angle, with bare arms and chest (except for the padding under the costume) holding the bear head under one arm to survey the landscape. He now seems out of place in the Arctic with his exposed skin, wind still howling, and no other human and animal life present (Fig. 3.1). Following montage logic, the film captures vivid details of the polar bear costume and the man's body in an unsteady chronology. Eventually the man in the polar bear suit, the Bear Man (as he is credited), performed by choreographer and dancer Robin Dingemans, leaves the Arctic and goes on a journey. We see the Bear Man walking along coastal paths, highways and across farmland, through a farm with cows, over a motorway bridge and through a disused industrial estate. These locations include: 'islands of Svalbard in the High Arctic to a kitchen in a house in London—via the beaches and headlands of Barra and Vatersay in the Outer Hebrides,

Fig. 3.1 Robin Dingemans in Fevered Sleep, *It's the Skin You're Living In*, 2013. (Film Still: David Harradine (director), courtesy of Fevered Sleep)

the M11 motorway, a dairy farm in Bedfordshire and the outskirts of Hackney and the Olympic Park' (Fevered Sleep n.d.-a). The figure in the costume appears in different states along the journey. At times he is wearing only the bottom half of the costume with large paws and furry legs, with the top half of his torso naked, but with the head of the polar bear on. The costume at times seems to cling to the white man; at other times he seems to be carrying it with him, until he sheds it completely at the end once he enters a terraced house in East London. He makes a cup of tea and curls up on the floor, skin exposed, as the camera finds his white facial hair in close-up, as if questioning the difference between the 'man' and the 'polar bear'. This film provides an affective metaphor for the way in which ecological relationships between humans and animals, between distant landscapes and home, between the everyday and climate change, are materially and performatively interconnected.

Fevered Sleep is a UK creative arts company that makes films, live performances, installations, books and digital arts, as a regularly funded Arts Council organisation. David Harradine and Sam Butler founded the company in 1996 and remain co-artistic directors. The film, part of Cape Farewell's Sea Change programme, was directed by Harradine and developed in conversation with Julie Doyle, researcher in media studies and climate change, during a residency at the University of Brighton.

The film offers a compelling reading of bioperformativity as it reframes images of climate change, revealing the performative nature of everyday objects and interactions, through a metaphorical conceit (man in a polar bear costume) and narrative agency. The film, through editing, framing, visual language and narrative, draws attention to the performativity of ecological relationships or what I am calling bioperformativity. This is a non-anthropocentric performativity of human-nonhuman assemblages, materially and symbolically.

I have previously written about *It's the Skin You're Living In* in terms of ecological anthropomorphism, a strategic kind of anthropomorphism that aims to draw out connections or isomorphisms between humans and more-than-humans (Woynarski 2015). Here I focus on bioperformativity and how it addresses the performative agency of human/more-than-human relations. Drawing on Jane Bennett's (2010, 2015) concept of ecomaterialism, bioperformativity can reveal the blurriness of the idea of the 'human' as separate from 'nature' through a narrative agency. For Bennett, the current anthropocentric thinking on the non-human as inanimate and/or instrumental prevents us from understanding the wide array of sensible forces, actants and matter that are operating and performing in relation to us. If all matter or 'things' are considered as having agency in some sense, the discursive differences between 'human', 'nature' and 'nonhuman' can be critiqued and questioned. Bioperformativity, as a way in which meaning is made in ecodramaturgies, disturbs strict and bounded anthropocentric hierarchies through recognition of the capacity for different kinds of more-than-human agency. *It's the Skin You're Living In* enacts bioperformativity by acknowledging the way in which humans/animals/climate are interconnected and tangled with murky boundaries between them.

In one sequence in the film, the Bear Man is shot from below, framed by power lines behind him. He moves his body fluidly, arms animated, chest bare in the looming shadow of the power lines. This sequence exposes the bioperformativity of the power grid, which is not a machine with fixed, predictable parts as a power blackout clearly illustrates. Revealing the lively elements of the power grid helps us to see all the actants we are operating in and with. For example, Bennett articulates the power grid as 'a volatile mix of coal, sweat, electromagnetic fields, computer programs, electron streams, profit motives, heat, lifestyles, nuclear fuel, plastic, fantasies of mastery, static, legislation, water, economic theory, wire, and wood' (Bennett 2010: 25). These actants have agency; however, it is only when something goes wrong, such as a blackout, that the multiple sites and loci of distributed agency become clear. The Bear Man and the power lines also point to the impact of power generation, particularly the amount consumed in the Global North (to support many energy-rich lifestyles as well as industry), and the resource extraction and subsequent oppression of people in the so-called sacrifice zone (such as Indigenous peoples and the Alberta Tar sands, the Dakota Access Pipeline route across traditional tribal land, the people living in the Niger Delta, to name a few). The film gestures to these connections by beginning the Bear Man journey in the Arctic and eventually ending up in London. The unusual and uncanny sight of the Bear Man in front of the power lines helps us to see the uncanny agency in the power grid, and gesture to the actants that make it up and the devastating ecological and social effects it produces (such as extractivism, environmental racism, high consumption, privatisation, profit imperatives and air pollution). *It's the Skin You're Living In* exposes this vibrancy and agency of the more-than-human, helping us understand the way in which we are interrelated to climate, polar bears, power lines and dairy farms—we produce ecological effects in inter-acting assemblages with these 'things'. Understanding these co-constituted effects may help us to see that we are all vital matter acting in the world.

In addition to the compositional dramaturgy, the editing of *It's the Skin You're Living In* is used to draw out how 'things' are related, despite how they are constructed as bounded and distinct in current social and political formations. The use of cuts between images, sometimes in split screen, uses a montage logic to draw connections and separations between places such as the Arctic and London and 'things' like power lines and the polar bear. Landscapes are spread across two images on screen, with the Bear Man walking across them. These 'things' are connected by ecological relationships where local actions and ideologies have far-reaching effects and consequences. 'We' are connected to the polar bear, even causing their 'migration' because of the capitalist, extractivist structures we live embedded in. The visual language of the film is described by Fevered Sleep as 'one of broken images, repeated actions and walking, walking, walking; a strange, sad and funny meditation on being human and being animal, lost in a changing world' (Fevered Sleep n.d.-a). This visual language also includes close-up of objects, highlighting their embodied ecological relationships. Towards the end of the film, once the Bear Man enters the house in London without the bear costume, we see a close-up of a bottle of milk being poured into a cup of tea. This draws connections back to an earlier sequence on a dairy/cattle farm, with the Bear Man walking past fenced-in cows. The cows stretch out their necks between the fencing, seemingly to meet the Bear Man as he walks by. He stops for moment and in split screen we see close-ups of his graceful bare arm movements as the camera pans the fencing surrounding him. The effect is that the Bear Man becomes briefly fenced in, echoing the position of the cows. A cup of tea with milk is an embodied connection to an animal but also to an ideology of industrial farming and likely exploitation of animals as well as the contribution of dairy and cattle farming to climate change-causing methane and CO_2 emissions. The political and ecological effects of this relationship are alluded to through the bioperformativity of the milk.

The reference to popular images of climate change in the film, including polar bears and the Arctic, enact an agency that is narratival and representational. Iovino and Oppermann position this agency in a narrative form as 'things' have stories: 'In material-ecocritical terms, the human agency meets the narrative agency of matter halfway, generating material-discursive phenomena in the forms of literature and other cultural creations' (2014: 9). I extend this narrative agency to theatre and performance analysis and ways of making meaning that constitute ecodramaturgies. Bioperformativity gives a name to agency as performative material that generates narrative effects.

The bear costume, in *It's the Skin You're Living In*, is one of the things with narrative agency. Not only with the ecological effects of its production and the material, labour and resources needed to make it but also as a communicative and indexical image within the film. One of the final images of the film is the camera panning across the Arctic landscape with the discarded polar bear suit lying on the ground, in isolation. For theatre scholar Carl Lavery, this image points to both the declining number of polar bears due to the effects of climate change as well as a future without humans: 'Whatever way one looks at it, this image is a pathetic image, an image of catastrophe, resonating with the death of a species…the image of the discarded suit appears to point forwards to a future where human beings, too, will have disappeared from the planet' (2014: 536). The costume then becomes 'an archival object for a humanity that will no longer be there' (Lavery 537). The bear suit has performative agency, within the film, its meaning is co-constituted with the man to the point that at the end of the film the man looks incomplete and bare without it and the costume looks deflated and lifeless, as if missing the man, through its materiality and narrative gesturing to a larger story of human extinction.

The film, through lingering close-ups, makes connections between 'fur and skin' finding similarities between them. These images are performative acts, as they draw attention to our living skin, or as Morton reminds us, 'we have others—rather, others have us—literally under our skin' (Morton 2010: 274). Polar bears in the Arctic have humans, particularly in the Global North, under their skin because of the way toxins travel through the atmosphere, accumulating in the Arctic and in the food chain, ingested by polar bears and also by humans. In the next chapter, I discuss how this bioaccumulation of toxins hits Indigenous women first, and hardest. The porosity of skin, the way it absorbs what is around it and excretes what we ingest, is an example of the way in which 'humanness' is not as strictly bounded as has been suggested by Western, anthropocentric ideology. If we are absorbing oxygen from the world around us, where do we end and the world 'outside us' begin? These kinds of distinctions become difficult to support as human bodies are made up of an assemblage of minerals, elements, bacteria, biota and other living things.

The film deconstructs and plays with the narrative agency of the iconic mascot of climate change, deliberately creating images and metaphors that resist the idea that the polar bear is a most visible (but distanced) victim of climate change. However, as Julie Doyle points out, the 'icons' of climate change are actually marking an erasure of peoples and locations. 'Since achieving iconic status as the "poster child" of climate change, the polar bear has diverted attention away from the impacts of climate change upon humans, particularly the cultures and livelihoods of the indigenous peoples of the Arctic and sub-Arctic' (2011: 53). Charismatic megafauna, mammals with large eyes, can be easily anthropomorphised, which may explain their role of 'poster child'. Yet it is Indigenous peoples, such as Saami, Aleut, Yupik and Iñupiat, Inuvialuit and Kalaallit in the Arctic, who are disproportionately suffering the effects of climate changes (despite their relatively small contribution to it) and their struggles and loss of habitat, tradition and culture are again erased. They are continually oppressed by the dominant power structures that silence their voices and sovereignty and designate them as inconsequential, or a box to be ticked, or at worst, 'primitive' and already extinct. This oppression is based in a history of cultural genocide and erasure with residential schools, forced adoption, destruction of tribal government forms, and contemporarily through lack of access to basic amenities such as clean drinking water, health and social care and political agency. In Chap. 6, I discuss in detail decolonised ecologies and how Indigenous works are creating ecodramaturgies that speak to their own lived experience and forms of knowledge. The dominant visual imagery of climate change performs emotional responses and signification as well as erasure and continuing colonisation.

As part of the collaborative work between director Harradine and Doyle, they produced a manifesto for making art about climate change, which is at once playful, prescriptive and open to possibilities that reimagine the conventional images of climate change. The process of making *It's the Skin You're Living In* reveals the ecological thinking underpinning ecodramaturgies. The manifesto includes statements such as 'It must not be literal', 'It must make me feel something', 'It must include the human', 'It must not contain images in which polar bears appear to be cute', 'It must be playful', 'It must be deadly serious' and 'It must not be singular' (Doyle and Harradine 2011). Part of the process stems from Doyle's assertion that the dominant visual language of climate change—'the polar bear, melted glaciers, flooded urban area, polar ice—have come so ubiquitous that they do little to foster a move active or embodied engagement with this issue' (2011: 145). Theatre and performance forms have the potential to facilitate 'active or embodied engagement', to be playful and serious, to make you feel something, through reimagining the visual iconography of climate change without taking it out of context, and including the human and the relationship between the 'distant' effects and our everyday lives. The manifesto tenets of not being singular, considering the audience, appearing in more than one form and place and being mobile all have the potential to foster a different way of thinking about climate change. The ecodramaturgy of the film, through this process of subverting and reimagining dominant visuals, myths, narratives, temporal and spatial understandings and forms, enacts a narrative agency towards ecological ways of looking at these values, attitudes, politics and power structures.

In addition to the ecodramaturgical process, new temporal and spatial understandings are opened up in the film through the Bear Man's journey. The sense of displacement emerges through the film as the man-bear journeys further and further from the Arctic. Una Chaudhuri (1995) writes about geopathology as the problem of place and extends it to include non-human animals as zoopathology: 'a disease of the ties that bind humans to animals' (Chaudhuri 2012: 46). 'This disease has a history in which the arts and representation are deeply implicated' (46). In their icon status, polar bears have been represented as primarily examples of displacement due to the effects of climate change. As the glacier ice melts at increasing rates, their 'home' diminishes and disappears. They are forced to swim further looking for food, making their way closer to human dwellings in this quest, moving from one 'home' to another. Chaudhuri writes about the taxidermy/photography project *nanoq: flat out and bluesome* (2001–2006) by Bryndis Snaebjorndottir and Mark Wilson in which they sought out 34 taxidermy polar bears across the UK in homes, museums and a pub. They were photographed in situ, strikingly out of place in the foyer of a home, holding a basket of flowers, or positioned on all fours next to a curtained window or in storage surrounded by crates and boxes. Chaudhuri refers to these as 'the very definition of the uncanny, in its etymological sense of "the unhomelike": the oddly estranged, the strangely out-of-place' (2012: 47). The polar bear costume in *It's the Skin You're Living In* produces a similar sense of the uncanny and estranged. A man in a half polar bear suit walking through a dairy farm with cows staring him down produces the out-of-placeness. This strange uncanniness is what produces the performative effects of the film and a metaphorical resonance with displacement due to climate change.

Chaudhuri refers to this as 'the theatre of species' which 'brings the resources of performance to bear on what is arguably the most urgent task facing our species: to understand, so as to transform, our modes of habitation in a world we share intimately with millions of other species' (2012: 50). This transformation in understanding our intimate interconnections is at the heart of ecodramaturgies. In positioning these interconnections and relationships as bioperformativity, I aim to open up possibilities for other ways of understanding the more-the-human world in which we are always already immersed. Rather than seeing a polar bear as a 'cute' and tragic figure of climate catastrophe, we might see it as intimately interconnected to us, as a co-habitant of a shared world, not unlike us. Through the journey of the man in a bear suit, *It's the Skin You're Living In* disrupts and dissects rigid categories that separate and distance humans from the more-than-human within the context of climate change. As Doyle suggests, 'climate change poses fundamental questions about what it is to be a human within a rapidly changing climate and world' (2011: 8). To be human in this context is to be interconnected to polar bears in the Arctic, cows on a farm and even cars on a motorway in assemblages of agency. The film performs the ecological effects of all these things, problematising human exceptionalism.

The bioperformativity of the 'things' in the film enact their narrative agency to disrupt cliché images and reconfigure them as lively, nuanced and complex. As Iovino and Oppermann suggest, 'The world's material phenomena are knots in a vast network of agencies, which can be "read" and interpreted as forming narratives, stories….the stories of matter are everywhere: in the air we breathe, the food we eat, in the things and beings of this world, within and beyond the human realm' (Iovino and Oppermann 2014: 1). *It's the Skin You're Living In* provides bioperformative stories of matter, material and more-than-human actants, demonstrating the way in which they can be reimagined in more ecological ways. Queer ecology provides a framework for this way of thinking.

Queer Ecology

One of the ways ecodramaturgies can activate an ecological sensibility beyond the human is through queer ecological questioning of the binary-making practices that divide humans and non-humans. *It's the Skin You're Living In* performs this disruption by 'queering' the idea of the 'human'. Barad argues that nature can be read as having a queer performativity that disrupts ontological distinctions between human and 'other' since queer is about questioning binaries, including nature/culture (2011). Queer ecology is a necessary strand of ecological thinking that can problematise and disrupt the configuration of human/nonhuman and nature/culture in a dichotomous relationship, as well as the social and political formations that support those binaries that result in oppression of certain humans and more-than-humans.

Theatre scholar Wendy Arons identifies queer ecology as an approach to ecocriticism that resists dualistic thinking about humans and nature, drawing on the way in which 'nature' does not maintain strict boundaries between humans and non-human, thus making it queer (2012: 565). Queer ecology theory is based on the idea that nature is queer as it does not operate in an assumed heteronormative way. As scholars Mortimer-Sandilands and Erickson write, 'queer, then, is both a noun and verb' (2010: 5). At the intersection of queer theory, ecology and evolution, this strand of thinking calls for a radical reconception of the human relationship to the ecological world. In different nuances and iterations, queer ecology calls into question the presumptive divide between human and nature, and nature and culture, in ways that undermine the social formations of heteropatriarchy, in an attempt to catalyse a shift in perspective towards ecological equitable viewpoints or sensibilities.

It's the Skin You're Living In can be read as taking on a queer ecology position by problematising bodily distinctions between human and non-human. In the film, we see similarities between the man and bear suit and therefore between the man and bear. As the man walks down the motorway, his legs in the bear suit, his bare feet covered with the claws of the costume and his arms and torso free and naked, a striking image of bodily confusion is generated. This seeming hybridity speaks to the hybridity of all human bodies as assemblages of cells and bacteria, containing water, chalk, metal and minerals and begins to trouble the binary of an 'exclusively' human body. Queer ecology is useful in challenging what we may consider categorically human or normative, opening up ways of thinking about what we share with the more-than-human and the ways in which we are also more-than-human. The Bear Man is a human construction of a non-human, yet it also speaks to the way in which all humans share similarities with non-humans as assemblages of living elements. Therefore, strict binaries between human and non-human become difficult to support. The film also references some of the discursive and material practices that separate human and non-human into dualisms (such as the Bear Man walking past the cows), with detrimental ecological and social effects. When these strict boundaries or dualisms are in place, it is easy to see climate change as a non-human problem and therefore easily ignored. The film demonstrates the way humans and non-humans are connected, making climate change a human and more-than-human problem.

It is the intra-activity of the man, the costume and the locations (Arctic, coast, farm, highway, city) that generates meaning. The uncanniness produced by their co-constituting turns our attention to the boundary-making practices that differentiate and separate human from animal, human from nature and nature from culture. For Barad, phenomena, or matter, emerge through intra-activity (rather than interaction, which presumes a prior independent existence). Barad seeks to reposition the discursive and materialising practices that separate the categories of beings and matter including subject/object, nature/culture and human/non-human. She does not seek to imbue the inanimate with life, but to turn our attention to the way 'things' are rendered 'inanimate' (or lacking agency) through discursive practices, rather than any essential quality they may or may not possess (2003). The bioperformativity of the Bear Man calls attention to the material-discursive practices that uphold distinct boundaries and our responsibility in maintaining the binaries and categories through excluding bodies and other matter. This exclusion is an important touchstone for queer ecology as it gestures to the way in which humans, when conceived as a singularity, exclude many of those who do not fit the Platonic ideal of who gets to claim 'human'.

The strength of queer ecologies is in what Di Battista et al. refer to as their 'affirmatively perverse and polyvocal imagination' (2015: 5). We see this in the journey of the Bear Man, as it opens up multiple spaces and interactions that trouble the normative. The Bear Man encounters cows, while the live cows encounter a performatively hybrid human/polar bear. There are many connections to be made by these imaginative juxtapositions. A Bear Man walking along the motorway, out of place as he is passed by speeding cars. On a beach, walking along the shore, not agile but encumbered by the furry legs and paws covering his feet. Or in the Arctic, bare and naked skin exposed to howling winds. The juxtaposition of these images draws attention to the structures of normativity, particularly in relation to the ecological, by creating absurd, humorous and sad situations in which to put a hybrid man/polar bear. The density of the images in the film affirms the polyvocal quality of queer ecological imagining.

Similar to the ways in which ecofeminisms contend that systems of power that cause gender-based oppression are based on the same ideological foundations of environmental oppression, queer ecologies suggest that the violence of heteropatriarchy and heteronormativity is based on the same ideological foundations that bifurcate nature/culture (Yep 2003:19). According to Anglin, 'distinctions between natural and unnatural, living and nonliving, and productive and nonproductive serve to maintain heteronormative and gender normative constructs that prevent political, social, and ecological growth from advancing' (Anglin 2015: 342). There is a double bind in that nature is paradoxically used as justification (for heteronormativity) and used to push away (as base or 'wild') anything that does not fit that strict distinction. 'Queers' therefore are both 'othered' in a way that is 'closer to nature' and seen as 'against nature' (Gaard 2004: 26). As Gaard explains 'today, all those associated with nature and the erotic [constructed as a dualism against reason] continue to experience the impact of centuries of Western culture's colonisation, in our very bodies and in our daily lives' (Gaard 2004: 39). According to Gray, queer subjectivities contribute to ecological thinking. Systems of oppression are ecological in that they are 'emergent from interconnected relationships that are complex, variable, persistent and hierarchical' (Gray 2017: 138). This points to an intersectional interpretation of queer ecologies and the coalition building needed to address the systemic structures of inequality.

The Privileged (2014) by performance maker Jamal Harewood takes an intersectional approach to the 'bear man' hybrid. The performance game invites the audience to enter a polar bear enclosure for an intimate experience with one of these endangered species. Notes from the zookeeper are in numbered envelopes, each placed on one of the seats formed in a circle in a studio space. There is no theatrical lighting, sound or stagecraft. There is only a man in a cartoonish and flimsy polar bear costume sleeping in character as a polar bear. As the audience read aloud the zookeepers notes, we find out the polar bear is named Cuddles and we are instructed how to interact with him. Facts about polar bears are listed, including the fact that under their white fur they have black skin. Audience instructions include playing with Cuddles, which start out as playful games and get progressively more sinister. The audience is asked to remove parts of the costume (his paws and then fur), which he resists. As pieces of the costume are forcibly removed by compliant audience members, Cuddles is revealed to be a nude Harewood, a Black, male-presenting human. Then the audience are asked to force him to eat a bucket of fried chicken that has been delivered on the door, which the instructions eventually say needs to be taken from him before he 'makes himself sick'. Cuddles resists all along, which sometimes means the instructions are carried out with violence. As the instructions become more sinister and forceful, the performance game elements begin to break down as debates heat up and some people walk out. In the version I experienced at South Street Arts Centre (Reading, UK) in 2018, there were walk outs and heated arguments about what we as audience members should be doing (i.e. enforcing the oppressive instructions on the 'bear man', including racist stereotypes, or obstructing the momentum of the performance by not following them) with a strong desire to get to the end of the performance. Harewood remains silent and in character as Cuddles the polar bear the entire time. It is a thought-provoking and uncomfortable show by design. Harewood draws parallels between racial oppression and speciesism through offering up hegemonic power structures for the audience to grapple with. As the audience, we were compelled to face up to issues of white power and supremacy, and the way societal structures enable and enforce them. These structures are racist but they are also anthropocentric and heteropatriarchal and are therefore tied to ecological objectification through the same ideological forces. The polar bear, as an image, metaphor and species, is further nuanced, moving away from 'climate change mascot' to reveal the multitudes of meanings and connections, in an intersectional ecological way.

Queer ecologies expose the underlying discriminatory systems that shape our understanding of human–nature relationships in multiple ways. Or as Alaimo points out, '"Nature" and the "natural" have long been waged against homosexuals, as well as women, people of color, and indigenous peoples' (2010: 51). This is poignantly demonstrated in *The Privileged*, as well as how complicit 'we' are in these normative and oppressive structures. Queer ecology is more nuanced and complex than simply adding queerness to a long list of oppressions; therefore, the very foundational ideologies that prop up these divisions and dominations need to be examined and refigured from the perspective of intersectional ecologies. Performance experiences like *The Privileged*, through a playful and troubling reimagining of the polar bear, starkly expose the intimate ties between constructions of nature, race, anthropocentrism, colonialism and heteropatriarchy. Similarly, *Symphony for Absent Wildlife*, through the use of human performers as animals, calls attention to how these structures have erased and eradicated humans and more-than-humans.

Symphony for Absent Wildlife

Felt blankets, constructed into animal masks and jackets, performers making the birdcalls of missing species, performing on traditional Indigenous land: all of these 'things' have thing-power. They operate in human/nonhuman assemblages that include the material and matter of the 'things' but also the land, the audience, the forest and the history of colonial oppression.

Symphony for Absent Wildlife was an outdoor symphony performance commissioned for Nuit Blanche festival in Calgary, Alberta, Canada, in 2014. The night performance took place in the Olympic and Municipal Plaza in Calgary, which is traditional Ktunaxa, Métis, Tsuu T'ina and Niitsitapi (Blackfoot) land. Thin, bare tree trunks sourced from local forests were installed around a raised performance space in the plaza. The tropes of classical music concert were present as the performance began with 19 orchestra members entering the space in a single-file line, linked by their hand on the shoulder of the person in front of them. Each member was wearing a full head animal mask made of felt and identical tailcoats also made of felt from reclaimed army blankets. Their silent march to the stage has a sense of foreboding. They represented woodland spirits, and each mask was designed to 'recall the spirits of the once abundant wildlife across the Albertan plains: bison, moose, wapiti, wolves, grizzly bears, and eagles' (Studio Orta n.d.). The brown felt animal heads were intricately sculpted to include horns, beaks, snouts and feathers. They were constructed to represent the now absent animals of Alberta.

Members of the orchestra stood in neat rows behind black music stands while one by one they walked to the front of the performance space to collect one of the small handcrafted wooden whistles displayed on a black plinth in front of the conductor. Each member examined the whistle before placing it under their mask and blowing in it to produce different birdcalls. After the orchestra members had collected their whistles, the symphony of birdcalls came alive. This cacophony of 'avian chatter' (Studio Orta n.d.) included hooting owls, woodpeckers and higher pitched chirping: '*Symphony for Absent Wildlife* brings a fragment of the disappearing natural environment, its sounds and the diversity of its fauna, closer to the city' (Studio Orta n.d.). The conductor, an unmasked white woman, walks around the orchestra, instructing members through touch to pause and then rejoin the symphony building up an iteration of a dawn chorus. Around 14 minutes into the performance, the birdsong builds into a frenzied crescendo before being abruptly cut short by the conductor, leaving a sense of violent loss. The orchestra bows and silently exits the space the same way they entered with hands to shoulders in a linked chain. In a 2016 installation for the opening of the Attenborough Arts Centre in Leicester, UK, the orchestra musicians were ceramic sculptures and the score was reimaged as an immersive audio experience. This work decentres the human (particularly the white settler) to focus on the loss of Indigenous peoples and land and more-than-humans.

Symphony for Absent Wildlife worked to unsettle anthropocentrism in performance, aiming instead at redrawing relationships horizontally. Greg Garrard has proclaimed that 'much ecocriticism has taken for granted that its task is to overcome anthropocentrism' (2004: 176), which is based on oppression and exclusion and part of the ideology of colonisation, racism, misogyny and ableism. Laying bare and addressing these socio-cultural forces is also the task of ecodramaturgies. The current climate crisis, and the underlying capitalist, extractivist logic dominating the Global North means that all disciplines are facing an ecological imperative. Ecodramaturgies can acknowledge and critique the agency of the human and more-than-human toward developing a non-anthropocentric way of thinking about theatre and performance. Throughout this book I argue that ecodramaturgies can and should expose ways of thinking about ecological relationships such as the relationships between humans and more-than-human nature, everyday actions and 'distant' climate change effects and the power structures of anthropocentrism, capitalism and colonialism that undergird them. In this section, I build on my earlier interpretation of bioperformativity and propose it as a way of thinking critically about ecodramaturgies as non-anthropocentric. Bioperformativity, because of the immanent acknowledgement of the material agency and performativity of the more-than-human, can provide a theoretical proposal for laying bare and critiquing anthropocentrism, in a way that is mindful of intersectional ecologies.

Symphony for Absent Wildlife is a continuation of the ecological work Lucy + Jorge Orta have been making as Studio Orta. Their work spans performative dinner feasts in public spaces, gallery and museum installations, outdoor sculpture, reimagined fashion and habitats and contemporary performance. They are provoked by questions such as: 'How can art practice pave a new critical role, faced with the growing problems of this world? How can it erase the contradictions between formal aesthetics and social function? How can works of art empower and nurture constructive dialogue? What contribution can we as artists make to human and environmental sustainability?' (Orta and Orta 2011: 5). Writing about the mutual dependency of 'us' and 'nature', depicted in their *Amazonia Expedition Sketchbook* (2009–2010), Lucy Orta states 'we are more dependent on nature than nature is on us—our presence brings about nature's decline and human decline with it, unless we choose to change and find solutions to these local and global problems placing us within nature, not outside of it' (Orta 2011: 84). Seeing and understanding the destructive agency of some humans and the socio-cultural forces that enable them brings attention to how ecodramaturgies can reveal ecological relationships.

The animal mask and matching felt tailcoats and trousers in *Symphony for Absent Wildlife* is one dramaturgical way of reimagining ecological relationships by creating a slightly uncanny juxtaposition with the human hands holding the whistle. Like *It's the Skin You're Living In*, the configuration questioned the idea of the human through the hybrid human-with-animal-head producing birdsong. The piece references the absent, erased and perhaps forgotten human and more-than-human occupants of the land. There is an emphasis on the missing and extinct flora and fauna of Alberta, recuperating their memory through the symphony of their sounds. Through the whistles, a dawn chorus of pre-colonial times is imagined. The heads of the orchestra members are completely obscured by their felt animal heads, unsettling the distinctions between human/ more-than-human. For example, the mask of the moose head features large sculpted horns and a large nose, with minimal eye holes making the person wearing it indistinguishable. This decentring of the human references the questioning and critiquing anthropocentrism, which is not about collapsing all distinctions and differences as Barad suggests, it is: 'to understand the materialising effects of particular ways of drawing boundaries between "humans" and "nonhumans"' (2011: 123). The intervention of bioperformativity here can help to unpack the way in which things create material and performative effects, including binary-making practices. It is the 'othering' practices of anthropocentrism, such as racism, that need to be troubled, critiqued, resisted and rejected, as has been done in unsettling the idea of 'who is human?' in queer theory, crip theory, feminism, postcolonial studies and critical race studies.

Symphony for Absent Wildlife with its reference to Indigenous species, land and peoples and the violence of settler colonialism, asserts that the category of 'human' is not fixed, stable or naturalised. As Rosi Braidotti points out, 'not all of us can say, with any degree of certainty, that we have always been human, or that we are only that. Some of us are not even considered fully human now, let alone at previous moments of Western social, political and scientific history' (2013: 1). This is particularly true when referring to the 'human' as constructed in the Enlightenment tradition who is a citizen with rights and a property-owner, which was typically a wealthy, white man. This is usually the Platonic idea of Man which all things are measured against, and an unmarked neutral being, a reductive and singular idea of the human: 'At the start of it all there is He: the classical ideal of "Man", formulated first by Protagoras as "the measure of all things", later renewed in the Italian Renaissance as a universal model and represented in Leonardo da Vinci's Vitruvian Man' (Braidotti 2013: 13). The bioperformativity of the absent wildlife in the symphony can be read from an intersectional perspective, where, as Bennett writes, 'all bodies become more than mere objects', particularly those 'routinely made to suffer because they do not conform to a particular (Euro-American, bourgeois, theocentric, or other) model of personhood' (Bennett 2010: 13). The symphony gestures not only to extinct animals but to missing traditional lands and Indigenous peoples who have been subjected to cultural genocide (e.g. through Indigenous children being forced into residential schools designed to erase Indigenous cultures, languages and religions through abuse and violence from 1880–1996). These practices point to the fact that Indigenous peoples did not get to occupy the category of 'human' in the eyes of Canadian government or Christian church. By acknowledging these structures that draw the boundaries of who gets to be 'human', the dominant image of the white, male, liberal subject is displaced with a more expansive idea of the human, always in relation to the lively matter of all 'things'.

Indigenous peoples were not considered 'human' in terms of citizen rights by colonising Europeans as alluded to in *Symphony for Absent Wildlife*. The choice of masks and costumes made of reclaimed felt blankets points to the settler colonialism of Indigenous Turtle Island, or Canada: 'The felt blanket represents an element of exchange between First Nation communities and the first European traders and settlers, and it became a "democratic" item of clothing' (Studio Orta n.d.). In the eighteenth century, letters between British forces officers evidence the intentional exposure of First Nations communities to smallpox through the distributing of infected blankets (Thornton 1990: 78) as part of a sustained genocide. The blankets carry this history and the impact of this intentional killing of Indigenous peoples by colonisers. The blankets are not about an equitable 'exchange' or a democratic item of clothing between settlers and Indigenous peoples, but rather an object of a war waged against them. The effects of the violence of colonisation are far reaching and are still being felt today with the oppression of Indigenous peoples in North America and the slow recovery from cultural genocide. Felt, in the performance, enacts a bioperformativity as it connects and reveals the oppressive structures that have delimited humans and their material effects.

Felt also features prominently in Beuys' work, including wrapping himself in large felt blankets for *I Like America and America Likes Me* (1974) in which he spent three days living in a room with a live coyote (named Little John) in New York. After arriving at the airport, he was put on a stretcher, wrapped in the felt and taken to the René Block gallery, never touching American soil. He used the felt to create different kinds of shapes and sculptures for his interactions with the coyote. The coyote is an 'animal-symbol' for America, with associations of 'wildness' and 'the untamed' (Pefanis 2018: 180). It also occupies a prominent place in many Indigenous mythologies in North America, which gives the felt a similar resonance to *Symphony for Absent Wildlife*. Beuys positions himself as a patient in an ambulance in need of healing from the coyote with the felt blanket representing insulation and protection. However, he operated more like a coloniser: trapping and caging a coyote and attempting to 'train' him to interact with Beuys. The coyote even ripped up one of the felt blankets. If the coyote stood in for a 'wild nature', Beuys captured and dislocated it. The interconnection of ecological violence and colonial violence can be read in this piece and *Symphony for Absent Wildlife* through the 'thing-power' of the felt blankets. They operate in assemblages that tie together the violence of colonialism, settler anthropocentrism and exclusionary ideas of the human, and the resulting climate and ecological emergencies as an outcome of these ideologies and social practices. The absence felt in *Symphony for Absent Wildlife* is not only flora and fauna once present and thriving in Alberta but also the original peoples of the lands[5] whose language and culture as well as livelihoods were decimated by settlers' claims on the land as a resource for their 'progress'.

Studio Orta's ecodramaturgy decentres the human as a singularity (the citizen, the white, liberal male subject) in favour of flora and fauna, with a symphony of birdsong rather than instrumental music, citing the absent Indigenous peoples (who were and are denied citizen rights). Humans have differentiated and asymmetrical agency, which needs to be recognised along with the agency of the more-than-human. Not all 'humans' bear the same responsibility or have the political or social capital to be able to respond. The bioperformativity of material agency, in the felt blankets as masks and costumes, the performers and the whistles of absent birds and species, highlights the socio-political-ecological practices that delimit and divide humans and more-than-humans.

The Weather Factory

Like felt animal masks, weather performs and is performative. Weather is a conversational lubricant and its unpredictability is often cited as evidence of climate change (see, e.g. *#crazyweather* in Chap. 5). It is also more than a neutral or natural backdrop for human action; it is understood and shaped by social, political and cultural forces. Weather is directly experienced but often difficult to conceptualise in terms beyond the human. Fevered Sleep's performative installation *The Weather Factory* (2010) was commissioned for the first season of National Theatre Wales, which is known as 'a theatre without walls' because it does not have a central base or theatre building, instead making work throughout Wales. Over the course of a year, director David Harradine interviewed local residents in

Penygroes, Wales, about the weather and came to think of it not only as 'a natural phenomenon—an effort of the relationships between the sea, the sun and the landscape—but also a human drama, a way for us to talk about ourselves and the places in which we live' (Fevered Sleep n.d.-b). Penygroes is a small village near Snowdonia National Park, which is home to Mount Snowdon, Wales' highest mountain and popular hiking spot. The performance took place in a house in Penygroes, rather than a theatre space. After collecting the key and directions from the Goat Pub, a small group of audience members (between 1–6 people) would tour the house. There was an absence of 'human' performers, and audiences were free to tour the house as they liked. Upon entering, audiences were met with a note in the front room, from the owner of the house, that invites them to 'make themselves at home' and go through any door that isn't locked. As it was close to Christmas, they were also invited to have a glass of sherry and mince pie, as an illuminated Christmas tree sat behind them. The front room was filled with dark wooden furniture of the kind typical of 1950s homes in the UK. Glass fronted cabinets were filled with crystal glasses with water (likely collected rainwater) and pieces of slate covering the top, tagged with numbers and symbols. A large, vintage radio played *Stormy Weather* by Etta James as the sheet music laid open on a piano in the corner. On the dining table a large map of the area was set up along with dice with symbols representing different weather. Barometers hung on the wall in the entryway.

As the audience makes their way through the house, they encounter one room that is ostensibly set up as an office with photos of a local landscape in different weather states covering the walls in rows. When the drawers of the desk or cabinets open they are full of meteorology instruments and mini projections of clouds and skyscapes. Another reception room contains stacks of televisions, each featuring a shifting skyscape with clouds slowly moving and changing. The cosy, domestic nature of the experience is disrupted when the door to the kitchen is opened and a room full of wind is revealed, with pots and pans shaking noisily on the countertop. It is raining in the basement, with large drops hitting a stone floor and crystal glasses on a small table set up to collect it. The rain is rhythmic and steady. Upstairs, there is a room full of hanging glass pendant lights, creating the soft, warm glow of twilight. Another room contains mirrors bouncing a light between them. The bathroom is covered in growing, squishy moss, engulfing the bath, walls and toilet. Fog fills another room, with an illuminated hanging light in the centre. Fevered

Sleep describe the piece as not only containing rooms full of weather but also landscapes of the region (Fevered Sleep n.d.-b).

Through an ecodramaturgy of visual imagery, materiality and experience of imaginative possibilities, we start to grasp the immense scale and pervasiveness of more-than-human agency and the intricate interconnections that constitute ecological relationships. *The Weather Factory,* by imitating weather events, enacts and performs 'weathering' (Neimanis and Hamilton 2018; Neimanis and Walker 2014): 'Weathering, then, is a particular way of understanding how bodies, places and the weather are all inter-implicated in our climate-changing world. Weathering describes socially, culturally, politically and materially differentiated bodies in relation to the materiality of place, across a thickness of historical, geological and climatological time' (Neimanis and Hamilton 2018: 80–81). Weathering is understanding that places, people and weather are materially related and interconnected. The way people speak about the weather is tied to an identity of place, but, more than that, weathering acknowledges that we are weather; we are not at a remove from the material effects of the world, particularly in relation to climate change. Weathering questions the distinction between weather and climate, as it is 'more-than-meteorological' (Neimanis and Hamilton 2018: 82), it is a description of social, cultural and political practices that expose some bodies (human and more-than-human) to the effects of the changing climate over others.

The Weather Factory literally brings home the weather, it also references measuring and cataloguing it, historicising it in the context of climate change. Within climate change-related fields of study, there has been a push to maintain a strict distinction between weather and climate. Weather is the changing, atmospheric conditions, directly experienced, while climate is the accumulation of weather patterns over time. However, some scholars have been questioning the usefulness of upholding this strict distinction. Mike Hulme argues for the need for a more inclusive and capacious understanding of weather and climate (2015). Scientific definitions of climate, such as the one offered by the World Meteorological Organization as a 'statistical description' do not get to the cultural, social, political, ethical and economic aspects of climate change (Hulme 2015):

> such definitions do not do justice to the deep material and symbolic interactions which occur between weather and cultures in places, interactions which, I believe, are central to the idea of climate. They too easily maintain a false separation between a physical world (to be understood through

> scientific inquiry) and an imaginative one (to be understood through meaningful narratives or human practices). (Hulme 2015: 176)

The Weather Factory unsettles that false separation by bringing the physical weather world into a symbolic and imaginative narrative based on human home-making practices. The weather is inside the home, situated in different rooms and overtaking at times the human domestic sphere. This resonates with feminist ecomaterialist weathering, where the distinction between weather and climate is necessarily porous and embodied according to Neimanis and Walker: 'attention to the material archive of weather in any body—a human, a starfish, a tropical storm—reveals the history of a lightning flash, or the thick presence of a February heat wave... Our aim is to reduce the distance between the enormity of climate change and the immediacy of our own flesh' (Neimanis and Walker 2014: 562). By reducing this distance and understanding ourselves as weather bodies (alongside more-than-human weather bodies), climate change becomes an embodied phenomenon, bringing it from a distant, observable pattern, to intimate everyday life. The house in *The Weather Factory* reduces the distance and prompts us to see the changing climate as our 'home' and inherent in the actions of everyday life—the weather is with us and within us. Researcher Lesley Duxbury further explicates the way weather and climate are always embodied: 'The weather has long been associated with the way we feel; breathing in and breathing out are taken for granted, and yet this same air is from the atmosphere that surrounds us containing all the extremes of the climate and inevitably becomes a part of us all...We are inseparable from weather and climate; we create it and participate in it.' (2010: 295). Reconfiguring weather as an assemblage of performative agency as in *The Weather Factory*, in which we, as humans, make and embody, destabilises the perceived distinction between us and the weather/climate.

As Doyle asserts earlier in this chapter, the popular imaginary and icons of climate change distance the experience of climate change, removing it from the 'everyday' of those living in the Global North, as well as erasing Indigenous and other bodies of colour that are struggling with the violent effects of climate change. Neimanis and Walker 'propose to bridge the distance of abstraction by bringing climate change *home*. As described in many climate change appeals, this home is a Western, urban, and domesticated home that more often than not seeks to extract itself from the weather-world. But we recall, too, that *oikos* is both "home" *and* another way of saying "eco"' (2014: 559). To bring home climate change, a more expansive understanding of home is needed, which I argue in more detail in Chap. 5, which is 'at once as distant as that melting icecap, and as close as our own skin' (Neimanis and Walker 2014: 559), with echoes of *It's the Skin You're Living In*. Neimanis and Walker propose reimagining climate change 'and the fleshy, damp immediacy of our own embodied existences as intimately imbricated' (559). Human action does not take place separately from climate and weather; it is in us and of us as *The Weather Factory* gestures to. Weathering is a shifting away from dominant climate change discourses that ask '"what should we do to stop climate change?" And instead asks "how is climate change me?"' (Neimanis and Walker 561). *The Weather Factory* brings the weather and climate home by facilitating an embodied experience of weather in a domestic account.

The rooms of *The Weather Factory* start off cosy with the front room playing *Stormy Weather* and another front room full of televisions playing skyscapes. These rooms are serene and contemplative. The noise picks up in the office and violence of the weather is revealed in the kitchen with the shaking pots and pans. The insistent rain in the basement has a kind of slow violence of erosion and sogginess. These different weather states gesture to the bioperformativity of the weather, and the way in which it is not a 'natural backdrop' but constitutive phenomena, made meaningful through intra-actions with humans and more-than-humans, in social, political and climatic interwoven formations. The domesticated home is disrupted through these at times of violence and intrusive weather states. *The Weather Factory* performs the idea that the weather does not exist in a vacuum; it is not an 'out there' phenomenon that can be defended against with double glazing. It is always already 'inside' and with us, an active co-participant. As Neimanis and Hamilton suggest 'this understanding of weathering also asks that we expand how we understand "the weather". Weather is pervasive in ways that makes distinctions between the meteorological and the social rather leaky, not unlike the much-critiqued nature/culture divide' (Neimanis and Hamilton 2018: 81). The weather is an embodied relationship that shapes the different contexts in which we are living. Neimanis and Hamilton also point out that weathering draws on intersectional feminism and politics of difference in that it is situated and 'not all bodies weather the same' (81). The house of *The Weather Factory* references a cosy, middle-class home in a village in Wales. The difference in weathering bodies is not immediate in this context when home is considered to be a three-bed family house. Although it does not acknowledge the differentiated ways in which weathering exposes bodies, the home is not immune to weathering through class or geography—the weather comes into regardless of where you live.

One of the ways to acknowledge the different weathering conditions is to bring the idea of total climate, in social, political and environmental terms, to the fore. To do this, Neimanis and Hamilton draw on Christina Sharpe's work on weather as a 'total climate' of anti-Blackness, as discussed in Chap. 2 in Thompson's *Salt*. For Sharpe, 'the weather is the totality of our environments; the weather is the total climate; and that climate is antiblack' (2016: 104). For Sharpe 'antiblackness is pervasive as climate. The weather necessitates changeability and improvisation; it is the atmospheric condition of time and place; it produces new ecologies' (106). Sharpe is thinking about what it takes to navigate these ecologies as a Black body, a Black person within a climate of anti-Blackness, where Black death is pervasive. Climate change and the weather conditions of climate chaos are interconnected with the social, historical, political, colonial climate. To link Sharpe's weather to Niemanis and Hamilton's weathering is not to appropriate or decontextualise Sharpe, but to acknowledge the intersectional, expansive idea of weather, the way in which it is 'more than meteorological', it is social, political, lived, historical and present:

> In the face of the greatest climatic transformation that human bodies have ever known, weathering means learning to live with the changing conditions of rainfall, drought, heat, thaw and storm as never separable from the "total climate" of the social, political and cultural existence of bodies. This includes anti-blackness, but also, we suggest, coloniality, misogyny and the resourcing and thingafication of other bodies—poor, queer, non-human, disabled. (Neimanis and Hamilton 2018: 82)

A total climate made up of ideologies and practices of racism, colonialism, xenophobia and heteropatriarchy means we need to weather the unprecedented conditions of a chaotic and changing climate. In the *Mothers of Invention* podcast (2018) climate justice campaigner Yvette Abrahams characterises how women South African farmers understand climate change 'as the white man stole the weather', which explained the changes they had been directly experiencing. She contends that because global warming is escalating about twice as fast in Africa as in the rest of the world, the effects cannot be denied or ignored, including drought and water rationing. Mary Robinson, former President of Ireland and co-host of *Mothers of Invention*, takes this explanation further, articulating the ways in which the 'white man' stole the weather by denial, fossil fuel investments and the gendered dimensions of climate change, including food responsibility and insecurity, leading to severe water shortages throughout Africa (Robinson and Higgins 2018).

By bringing the weather inside, ideas of shelter are called into question in *The Weather Factory*. What happens to the defences we enact to keep the outside world out if the weather is inside? This becomes another 'leaky' boundary. To weather, as in the case of *The Weather Factory*, is to understand the way in which different bodies are exposed to weather in differentiating ways. Not everyone enjoys the cosiness of a centrally heated home and more and more, the energy required to maintain this is having disproportionately negative effects on the Global South. Rising costs of energy also mean that older people and poorer people's access is being limited. Exposure to elements through damp, leaks and draughts, or rough sleeping, is an everyday reality for many people. Weathering is an embodied relationship as this exposure affects health and well-being. In *The Weather Factory*, the audience gets to leave; their exposure is limited to 45 minutes with a glass of sherry and a mince pie. As Neimanis and Hamilton argue, 'Weathering reminds white settler colonial bodies that learning to weather better cannot be about fortifying our own havens; weathering better requires interrupting our existing patterns of weather-making, broadly construed' (2018: 82). The violence at times of the weather in the house of *The Weather Factory* interrupts a domesticated safety of weathermaking but it does not draw attention to the differences in weathering and the oppression structures behind it. For Neimanis and Hamilton, weathering needs to address the ways in which our existing patterns of weathermaking are destructive. Weathering the social, political and geophysical climate cannot mean a redoubling of neocolonial anthropocentrism and neoliberal individualism, as perhaps suggested by the cosy scenography of *The Weather Factory*. Weathering better, in intersectional ecologies, means to understand how these ideologies shape our understanding of weather and broader ecological relationships, acknowledging the violence inherent in some of these relationships and trying to stop some of that damage. We need to weather as praxis, understanding our agency within our everyday, domestic lives, and then remaking our weather-worlds or total climate of white, heteropatriarchy in which ecological oppressions are gendered, raced and classed. *The Weather Factory* thinks beyond restrictive limits of anthropocentrism and acknowledges a more-than-human weathering.

The performances discussed in this chapter uncover the ecologically material and narrative, indexical agency of the more-than-human that can be read as reconfiguring human and more-than-human relations away from hierarchical verticality. We share qualities and ecological interconnectedness with trees, polar bears, missing species and the weather, creating material effects that act on/with each other in dynamic relationships of ecology, or bioperformativity. Ecodramaturgies manifest and complicate this ecological entanglement. The current ecological situation has created an imperative for theatre and performance to think beyond the human, to the structures that prop up anthropocentrism, heteropatriarchy, coloniality and racism in order to enact more ecological modes of being-in-the-world.

Notes

1. With the use of the 'performativity' I am not suggesting a dichotomous relationship with theatricality or that theatre spaces cannot enact an ecomaterialism. As Chaudhuri and Enelow have argued, the theatrical black box can be a form for performing Bennett's ecomaterialism (2014).
2. I first began to think about *7000 Oaks* when I collaborated on conference paper for the American Society of Theatre Research in 2012 with Meghan Moe Beitiks and Bronwyn Preece entitled: 'Restorations, Actions, Ecoventions and Questions: An exploration of the potential of ecologically restorative flash mobs'. I thank them for their collaboration.
3. See Alston (2013), White (2013), Bishop (2012).
4. Monoculture tree planting programmes can cause more carbon to be emitted in the atmosphere and other unintended ecological effects (Elbein 2019). For example, a large-scale forest fire in Fort McMurray, Alberta, Canada, revealed the pitfalls of turning wetlands into forests, as the trees sucked up the groundwater, intensifying the fire.
5. This is not to ignore the many Indigenous nations currently in Alberta, or code them as 'historical'. I explore some contemporary Indigenous thinking and activist movements in Chap. 6 in relation to the reductive idea of 'humans' as responsible for the Anthropocene.

References

Alaimo, Stacy. 2010. Eluding Capture: The Science, Culture and Pleasure of 'Queer' Animals. In *Queer Ecologies: Sex, Nature, Politics, Desire*, ed. Catriona Mortimer-Sandilands and Bruce Erickson, 51–72. Bloomington: Indiana University Press.

Alston, Adam. 2013. Audience Participation and Neoliberal Value: Risk, Agency and Responsibility in Immersive Theatre. *Performance Research* 18 (2): 128–138.

Anglin, Sallie. 2015. Generative Motion: Queer Ecology and Avatar. *The Journal of Popular Culture* 48 (2): 341–354.

Arons, Wendy. 2012. Queer Ecology/Contemporary Plays. *Theatre Journal* 64 (4): 565–582.

Barad, Karen. 2003. Posthumanist Performativity: Toward an Understanding of How Matter Comes to Matter. *Signs: Journal of Women in Culture and Society* 28 (3): 801–831.

———. 2011. Nature's Queer Performativity. *Qui Parle* 19 (2): 121–158.

Beitiks, Meghan Moe. 2018. The Plant Is Present. In *Botanical Speculations: Plants in Contemporary Art*, ed. Giovanni Aloi, 63–67. Newcastle upon Tyne: Cambridge Scholars Publishing.

Bennett, Jane. 2010. *Vibrant Matter: A Political Ecology of Things*. Durham: Duke University Press.

———. 2015. Systems and Things: On Vital Materialism and Object-Oriented Philosophy. In *The Nonhuman Turn*, 223. Minneapolis: University of Minnesota Press.

Beuys, Joseph. 2012. 7,000 Oaks: Conversation with Richard Demarco//1982. In *Nature*, ed. Jeffrey Kastner, 167–170. London; Cambridge, MA: Whitechapel Art Gallery; MIT Press.

Beuys, Joseph, Johannes Stuttgen, Elisabeth Huhn, and Chistoph Stuttgen. 1982. *Portrait of an Art Performance—Joseph Beuys 7000 Oaks*. Gelsenkirchen, Germany: FIU (Free International University).

Bishop, Claire. 2012. *Artificial Hells: Participatory Art and the Politics of Spectatorship*. London; New York: Verso Books.

Braidotti, Rosi. 2013. *The Posthuman*. Cambridge, MA: Polity.

Chaudhuri, Una. 1995. *Staging Place: The Geography of Modern Drama*. Ann Arbor: University of Michigan Press.

———. 2012. The Silence of the Polar Bears: Performing (Climate) Change in The Theatre of Species. In *Readings in Performance and Ecology*, ed. Wendy Arons and Theresa J. May, 45–57. London; New York: Palgrave Macmillan.

———. 2016. The Fifth Wall: Climate Change Dramaturgy. *Howl Round*, April 17. Accessed June 28, 2020. https://howlround.com/fifth-wall.

Chaudhuri, Una, and Shonni Enelow. 2014. *Research Theatre, Climate Change, and the Ecocide Project: A Casebook*. New York: Palgrave Pivot.

Cohen, Jeffrey Jerome. 2013. *Prismatic Ecology: Ecotheory beyond Green*. Minneapolis: University of Minnesota Press.

Di Battista, Amanda, Oded Haas, and Darren Patrick. 2015. Conversations in Queer Ecologies: An Editorial. *UnderCurrents: Journal of Critical Environmental Studies* 19: 3–5.

Diamond, Elin, ed. 1996. *Performance and Cultural Politics*. London; New York: Routledge.

Dolphijn, Rick, and Iris van der Tuin. 2012. *New Materialism: Interviews & Cartographies*. Ann Arbor: Open Humanities Press.

Doyle, Julie. 2011. *Mediating Climate Change*, Environmental Sociology. Burlington, VT: Ashgate.

Doyle, Julie, and David Harradine. 2011. Manifesto for Producing Artwork on Climate Change. Accessed September 12, 2018 http://arts.brighton.ac.uk/projects/here-today-moving-images-of-climate-change/manifesto.

Duxbury, Lesley. 2010. A Change in the Climate: New Interpretations and Perceptions of Climate Change through Artistic Interventions and Representations. *Weather, Climate, and Society* 2 (4): 294–299. https://doi.org/10.1175/2010WCAS1053.1.

Elbein, Saul. 2019. Tree-Planting Programs Can Do More Harm than Good. *National Geographic*, April 26. https://www.nationalgeographic.com/environment/2019/04/how-to-regrow-forest-right-way-minimize-fire-water-use/.

Fevered Sleep. 2010. *The Weather Factory*. Performance/Installation. National Theatre Wales, December 7–24.

———. n.d.-a. It's the Skin You're Living In. Accessed April 10, 2019. http://www.feveredsleep.co.uk/current/its-the-skin/.

———. n.d.-b. The Weather Factory. Accessed April 08, 2019 https://www.feveredsleep.co.uk/project/the-weather-factory.

Gaard, Greta. 2004. Toward a Queer Ecofeminism. In *New Perspectives on Environmental Justice: Gender, Sexuality, and Activism*, ed. Rachel Stein, 21–44. New Brunswick; London: Rutgers University Press.

Garrard, Greg. 2004. *Ecocriticism*. Abingdon, Oxon; New York: Routledge.

Gray, Jonathan M. 2017. Heteronormativity without Nature: Toward a Queer Ecology. *QED: A Journal in GLBTQ Worldmaking* 4 (2): 137–142.

Harradine, David. 2013. *It's the Skin You're Living In*. Film. Fevered Sleep.

Hayman, Richard. 2003. *Trees: Woodlands and Western Civilization*. London: Hambledon and London.

Heddon, Deirdre. 2013. Pitching the Forest. Presentation. In *Performance and the Environment Symposium, Royal Central School of Speech and Drama*, London, January 11.

———. 2016. Confounding Ecospectations: Disappointment and Hope in the Forest. *Green Letters* 20 (3): 324–339.

Hulme, Mike. 2015. Climate. *Environmental Humanities* 6 (1): 175–178.

Iovino, Serenella, and Serpil Oppermann. 2014. *Material Ecocriticism*. Bloomington: Indiana University Press.

Kershaw, Baz. 2007. *Theatre Ecology: Environments and Performance Events*. Cambridge; New York: Cambridge University Press.

Lavery, Carl. 2014. Polar Bears, Climate Change and the Ethics of Limitrophy. *Contemporary Theatre Review* 24 (4): 535–536.

Mortimer-Sandilands, Catriona, and Bruce Erickson, eds. 2010. *Queer Ecologies: Sex, Nature, Politics, Desire*. Bloomington: Indiana University Press.

Morton, Timothy. 2010. Guest Column: Queer Ecology. *PMLA* 125 (2): 273–282. https://doi.org/10.1632/pmla.2010.125.2.273.

Neimanis, Astrida, and Jennifer Mae Hamilton. 2018. Weathering. *Feminist Review* 118 (1): 80–84. https://doi.org/10.1057/s41305-018-0097-8.

Neimanis, Astrida, and Rachel Loewen Walker. 2014. Weathering: Climate Change and the 'Thick Time' of Transcorporeality. *Hypatia* 29 (3): 558–575.

Nixon, Rob. 2011. *Slow Violence and the Environmentalism of the Poor*. Cambridge: Harvard University Press.

Orta, Lucy. 2011. *Operational Aesthetics, The Work of Lucy + Jorge Orta*. London: University of the Arts.

Orta, Lucy, and Jorge Orta. 2011. *Lucy + Jorge Orta: Food, Water, Life*. New York: Princeton Architectural Press.

Pefanis, George P. 2018. Is the "Animal-Event" Possible? Animal Precariousness and Moral Indeterminacy in Performance. *New Theatre Quarterly* 34 (2): 176–185.

Robinson, Mary, and Maeve Higgins. 2018. The White Man Stole the Weather. Mothers of Invention Series 1, Episode 2. Accessed April 3, 2019. https://www.mothersofinvention.online.

Sharpe, Christina Elizabeth. 2016. *In the Wake: On Blackness and Being*. Durham: Duke University Press.

Spaid, Sue. 2002. *Ecovention: Current Art to Transform Ecologies*. Cincinnati: Contemporary Arts Center.

Studio Orta. 2014. *Symphony for Absent Wildlife*. Performance, Nuit Blanche, Calgary, Canada, September 20.

———. n.d. Symphony for Absent Wildlife. Accessed April 3, 2019. https://www.studio-orta.com/en/artwork/469/Symphony-for-Absent-Wildlife-Spirits-of-Alberta.

Thornton, Russell. 1990. *American Indian Holocaust and Survival: a Population History Since 1492*. Norman: University of Oklahoma Press.

White, Gareth. 2013. *Audience Participation in Theatre: Aesthetics of the Invitation*. London; New York: Palgrave Macmillan.

World Wildlife Fund. 2014. Plant a Tree Seed, Save a Planet. Accessed January 03, 2014. http://wwf.panda.org/how_you_can_help/campaign/plant_seed/.

Woynarski, Lisa. 2015. A House of Weather and a Polar Bear Costume: Ecological Anthropomorphism in the Work of Fevered Sleep. *Performance Research* 20 (2): 24–32.

Yep, Gust A. 2003. The Violence of Heteronormativity in Communication Studies: Notes on Injury, Healing, and Queer World-Making. *Journal of Homosexuality* 45 (2–4): 11–59.

CHAPTER 4

Bodily Ecologies: Exposure, Participation and Immersion

In 2007, in a landmark article connecting ecocriticism to theatre and performance studies, Theresa J. May writes that a potent ecocriticism in theatre and performance 'will participate in the discourse of environmental justice by foregrounding the body as medium between material and metaphoric worlds, and mapping the connections between social injustice, human and other bodies, and environmental exploitation' (May 2007: 101). This means an ecocriticism that complicates and nuances reductive environmental discourses or exposes harmful ideologies. This chapter takes up May's call to foreground the body, as material and metaphor, mapping the intersections between 'social injustice, human and other bodies, and environmental exploitation', in Violeta Luna's *NK603: Action for Performer & e-Maiz* (2014a), Cherríe Moraga's *Heroes and Saints* (1992), Eeo Stubblefield's *Still Dance with Anna Halprin* (1997–2000), Deke Weaver's *Wolf* (2013) and Rimini Protokoll's *World Climate Change Conference* (2014). The body as a multiplicity is foregrounded across a range of different theatre and performance practices to theorise an expansive idea of bodies, open to humans and more-than-humans, recognising the material differences between them.

Performance is an embodied medium (Garner 1994), and in a sense, bodies sharing space can be one aspect that differentiates theatre and performance from other artistic mediums. Philosopher David Abram refers to the body as the material locality of our intersection with the living world (1997). It is this embodied experience of the more-than-human world

L. Woynarski, *Ecodramaturgies*, New Dramaturgies,
https://doi.org/10.1007/978-3-030-55853-6_4

that cultivates a sense of interconnection with it. For Abram, when we are attentive to our experience as affective bodies (rather than 'intangible minds'), we become sensitised to exchanges with the countless other bodies in the world and 'find ourselves alive in a listening, speaking world' (1997: 86). Our bodies are always already ecological and tied into the multiplex ecological relationships that structure the living world. Bringing this concept to bear on theatre and performance opens up ecodramaturgies to bodily practices of all kinds.

The focus of this chapter is how different bodies participate and are immersed and exposed by ecological issues and performance forms. The aesthetics (of ecodramaturgies) are based on making meaning from, according to Johnson, 'images, qualities, emotions and metaphors that are all rooted in the body's physical encounters with the world' (Johnson 2007: 17). Ecodramaturgies can engage the body as in and of ecological matter and material, drawing out new frames of thinking about bodily relationships and encounters. Ecodramaturgies draw on embodied practices that resonate and situate the performer/audience/participant within what Alaimo refers to as complex 'networks of risk, harm, culpability and responsibility' (Alaimo 2016: 3) as well as pleasure. My reading of the bodies in performance draws on the unequal and intersectional ways in which bodies are differentially exposed.

There is an activist slant to much of the work on ecological embodiment. For example, for Stacy Alaimo embodiment has a political and ethical dimension as 'to think as a body—indeed as a body that is part of the substantial interchanges, flows, and substances of the co-extensive world—is an entangled, provisional, highly mediated, but also potentially ethical and political endeavour' (2013: 16–17). She suggests that it is in the contact between humans and the more-than-human that political and ethical possibilities exist (Alaimo 2010: 2). Alaimo's theory of trans-corporeality, in which it is understood that humans are always already 'intermeshed' in the ecologically material world, highlights the inseparability of the human and the more-than-human world or 'the environment' (2010: 2). For me, this ethical and political edge requires a reformulation of bodies and behaviour related to the more-than-human based on how differentiated bodies are exposed differently, which ecodramaturgies can make meaningful.

In Chap. 3, I argued that bioperformativity provides a useful framework for thinking about more-than-human performativity, expanding the idea of who gets to be the subject or considered 'human'. Bodily ecologies encompass multiple bodies and acknowledge the differentiated power relations between them. Alaimo contends that ecomaterialism, with its insistence on matter and 'the substance of what was once called "nature", acts, interacts, and even intra-acts within, through, and around human bodies and practices' (2016: 1). Bodily practices and experiences in and of theatre and performance are a way to decentre human exceptionalism and the 'western man' as subject, expanding ideas of which bodies matter and how the 'matter' of bodies contains narratives.

Bodies, both human and non-human, provide an eloquent example of the way matter can be 'read as a text', as Iovino and Oppermann write:

> Whether performing their narratives as statues in a square, teachers in a classroom, plankton in the ocean, fossils trapped in a stone wall, or chickens in industrial factory farms, bodies are living texts that recount *naturalcultural* stories. (2014: 6)

This chapter is full of bodies as text, performing narratives as wolves, maize plants, poisoned and exposed farm workers, fallen trees and politicians negotiating a climate change agreement.

Embodiment as the locus of our relationship with the world means that our body is implicated in a number of ecological relationships. I think it is important to include a caveat here: bodily ecologies do not imply a body/mind dualism. Bodies are not machines controlled by 'ghosts' but thinking beings. Bodies contain minds allowing rationality, intellect and thinking to be experienced in the body rather than configured as separate from the mind in a dualism.

The discussion in this chapter focuses first on exposure and then on participation and immersion. Analysing the performances through these three related concepts illuminates different aspects of bodily ecologies, practices and engagements and allows for an expansive sense of social, political and material ecological interactions across very different performance forms. Exposure centres on the ecological body in Mexican performance artist Violeta Luna's work *NK603: Action for Performer & e-Maiz* (2014a) and Cherríe Moraga's *Heroes and Saints* (1992) to consider how environmental justice issues are felt and centred in different and marginalised bodies. *Still Dance with Anna Halprin* (1997–2000) is read with Ana Mendieta's *Siluetas Series* (1973–1980) and Ingrid Pollard's photography to further theorise the tension between bodies as 'natural', gendered, raced, aged and inscribed with neoliberal values. In participation and immersion, the focus is on Deke Weaver's *Wolf* (2013) and Rimini

Protokoll's *World Climate Change Conference* (2014), thinking about how ecological relationships can be revealed through performance with material, immersive encounters with human and more-than-human bodies. As Tim Ingold has written, 'the distinction between environment and nature corresponds to the difference in perspective between seeing ourselves as beings *within* a world and as beings *without* it' (2000: 20). In order to break apart binaries that conceptually separate humans from nature, we need to understand our relation as one of being within and of the material environment. Exposure, immersion and participation, or being *within* and of a world, in an embodied way, is necessary to move towards an understanding of ecology. Ecodramaturgical strategies can reveal and expose the way in which we are immersed within multiple ecological relationships, presenting a way of thinking about human relationships with the ecological world from an embodied perspective.

Exposure

NK603: Action for Performer & e-Maiz

Mexican performance artist Violeta Luna enters through the side of a dimly lit proscenium theatre auditorium through an audience entrance.[1] She is wearing a skirt made out of dried maize husks, a grey chal (shawl) and a straw hat with two long black plaits, and she is brandishing a machete. She moves slowly and silently through the rows of audience seating making defiant eye contact with audience members in the low lighting. She walks towards the raised stage and climbs up. Black and white images of Mexican farmers (*campesinos*), protests, scientific testing and maize plants are projected on a white sheet suspended behind her on stage, while a sound track of news reports and corporate videos about the safety of genetically modified (GM) foods plays. This is how *NK603: Action for Performer & e-Maiz* begins. As I sit in the auditorium of the theatre in Massey University in Wellington, New Zealand, I am shocked, angered, uncomfortable and compelled at different points by Luna's wordless performance. The work goes on to enact the links between exposed living bodies: her body, corn plants, farmer's bodies and livelihoods, political bodies and power.

Once on stage at the beginning of the performance, Luna kneels down and ties a white sheet around her nose and mouth. She does not speak throughout. She slowly peels an ear of corn and places it on a sheet in front of her. She then holds up a deep purple dried cob and places it on the sheet. She spits blood on the cobs, as if performing a ritual. Her upper body, bare once she removes the chal, is painted a deep purple representing the heart and sacredness in Maya cosmology, with seeds stuck to her chest forming the shape of a blossoming plant. She has a maize plant painted on her back, symbolising the inseparability of Indigenous cultures, corporeality and maize (Fig. 4.1).

In the performance, Luna puts on white medical gloves and tightly ties her long black plaits into a gag around her mouth. She takes a metal hooked surgical instrument and drags and tugs her head and neck in different directions. The voice in the soundtrack slows down and distorts while the score of electronic music builds in intensity. She uses medical tools to contort and prod her body as she takes surgical pliers and rips off

Fig. 4.1 Violeta Luna, *NK603: Action for Performer & e-Maiz*, 2014. (Photo: Greg Craig, courtesy of Violeta Luna)

the seeds stuck to her chest. A scientist in a white lab coat tightly binds her torso in duct tape. She puts on a metal corset with sharp nails projecting from it and injects her arm with a needle. She puts on a wired headpiece that forces her mouth open. She tries to speak but her words are incomprehensible and she has to swab her mouth with cotton. The manic video collages and electronic music (by longtime collaborators Roberto Varea and David Molina) violently intervene in the action. She takes a pouch with 'Made in USA' written on it and pours blue dough onto a tortilla press, forming a tortilla. Tortillas are made with ground maize and are a staple of the Mexican diet. She holds up the blue tortilla while the headpiece contorts her face into a maniacal smile. Near the end of the performance, the scientist cuts off the tape binding her torso and reveals that the painting of the maize on her back has transferred to the duct tape bonds. The rich tradition of growing maize has now been transferred to multinational corporate interests and the happy myth of progress that the bonds are standing in for. She then takes a red scarf, ties it around her nose and mouth, and stares defiantly out at the audience, a clear allusion to the Zapatista movement of the 1990s, in which Indigenous peoples and *campesinos* staged a revolution for agricultural and social reforms in Chiapas, Mexico.

Through affective imagery drawn from *campesinos*, anti-GMO protests, Mexican Indigenous rituals, the Zapatista movement and biotechnology experimentation, Luna performs the interconnectedness of human health and ecological health in her piece *NK603: Action for Performer & e-Maiz*. At once disturbing and compelling, Luna performs the manipulation of genetically modified (GM) maize crops on her body, connecting the maize plant (also known as corn) to the manipulation of her body and well-being. The title, NK603, refers to one strain of genetically modified maize introduced in Mexico against strong protests from farmers. NK603 is a Roundup® ready corn plant produced by Monsanto, a multinational biotech corporation and the largest producer of GM seeds[2] (Monsanto 2002). It is a maize hybrid that is resistant to the herbicide Roundup® (also produced by Monsanto), which means that maize crops can be sprayed with the herbicide and will not die. Luna's performance draws on maize as culture, heritage and a staple crop for Mexicans. The piece has toured festivals around the world since 2009, including Colombia, Serbia, US, Peru, Spain, New Zealand and Mexico, bringing the GM maize conflict to a global audience.

Many Indigenous (including Inuit, First Nations, Mesoamerican, Aboriginal and Maori) worldviews do not conceptualise a separation between an individual and the land on which they live. As Abram suggests, referring to Navajo cosmology, 'the health, balance, and well-being of each person is inseparable from the health and well-being of the enveloping earthly terrain' (1997: 237). Luna's work enacts this inextricable link between the earthly terrain and the health and well-being of Indigenous peoples, cultures, practices and livelihoods in the context of globalisation. Through her body, she connects maize growing, the impact of GM crops and land health to identity and Indigenous cultures. Historically, Indigenous Latin American cultures, as well as many First Nations and Indigenous nations of now North America,[3] have harvested maize with over 60 indigenous species grown. Culturally and spiritually, maize is foundational to the Maya creation story as it is considered the flesh and blood of the Maya people (Luna 2014b). Monsanto has been distributing GM maize (NK603) in Mexico since 2009, which it is currently only allowed to grow in an experimental form, although there is lobbying for full legalisation. The introduction of Monsanto and GM crops into maize agriculture will not only result in a loss of biodiversity, potentially making 'native' varieties extinct, but will have a profound impact on Indigenous people and culture (Luna 2014b). Luna locates this contested ecological site on her body in the performance; it is the site of memory, culture and capitalist intervention. The well-being of her body is inextricably connected to the land and its ecology. The body in performance, as May suggests, reveals ecological relationships: 'foregrounding the body also brings into focus the web of social, political, economic, and ecological systems that touch our bodies' (2007: 101). Luna's work reveals the ecological body as one of tension and conflict as well as connected to vital matter. In a sense, all of our bodies are ecological in that they are located in the world, are living things and are full of living things. As Bennett (2010) reminds us, our bodies are assemblages of vibrant material with thousands of microbes and bacteria living on and under our skin.

As Luna is poked and prodded, bruised and bloodied, the physical health of her body parallels the loss of biodiversity and genetic manipulation of the maize plants. Her body is implicated in the interventions into the land, in the globalisation of rural Mexico and in the loss of heritage and culture associated with maize. This work is also rooted in environmental justice, an activist gesture in response to the systemised inequality of neoliberal capitalism, oppression of the culture and heritage of

Indigenous people, their ongoing resistance, and the use of GM crops as monoculture. The performance does not represent land health; rather it performs the inseparability of ecological conditions and human well-being.

Luna creates an image and narrative that signals what Di Chiro describes as 'a future dystopic scenario where corporate agricultural biotechnology has infiltrated the domain of environmental and public health' (2004: 140). This brave new world is modelled on the multinational Monsanto Corporation's biotech agricultural practices stemming from the widespread use of their powerful herbicide Roundup® and the bioengineered crops produced by Monsanto which resist it, which include 'economically important plants such as canola, corn, soybeans, cotton, and sugar beets that are resistant to the company's widely used herbicide' (Di Chiro 141). NK603, the genetically modified maize, is economically as well as culturally and ecologically important. Monsanto promotes this as in line with sustainable agricultural practices (Di Chiro 141) while others argue it supports harmful industrial agricultural practices, adapting the crops to suit these practices, rather than changing destructive practices (Di Chiro 156). The World Health Organisation has named the active ingredient in Roundup®, glyphosate, as a probable carcinogen to humans (Stevens 2018). According to an investigation by *Le Monde* in 2017, new evidence of Roundup® causing cancer was covered up by Monsanto (Foucart and Horel). In August 2018, Dewayne Johnson, a groundskeeper for the California School District, won $289 million USD as a California court ruled that his terminal non-Hodgkin lymphoma was caused by his exposure to Roundup® in his job, although deemed safe by Monsanto.[4] Monsanto was acquired by pharmaceutical company Bayer in 2018 and plans to phase out the use of the name Monsanto. With Monsanto, Bayer is now one of the largest biotechnology pharmaceutical companies in the world.

Luna's work positions these biotech GM seeds as a threat to the way Indigenous cultures have grown maize in Mexico. It is thought that you do not cultivate maize; rather you have a conversation with it (Luna 2014b). Corn was a key factor in the thriving of Indigenous peoples in what is now known as Mesoamerica, the South-Central Andes and eastern North America before settler colonialism. The development of agriculture around corn, grown with beans and squash, including complex irrigation systems, supported a large population where according to Dunbar-Ortiz, 'people lived long and well with abundant ceremonial and recreational periods' (2014: 17), before the genocidal conventions of European 'discovery'. The relationship between corn and Mayans is foundational: 'the first great cultivators of corn were the Mayans… religion was constructed around this vital food' (Dunbar-Ortiz 18). Mayan influence was felt throughout Mesoamerica and corn agriculture spread to the Aztecs, and then through extensive trading networks to Pueblos, and then to 'Algonquin, Cherokee, and Muskogee (Creek) peoples of the eastern half of North America' (21), all of whom have traditional commemorative practices for 'the gift of the corn' (Dunbar-Ortiz 31). The introduction of Monsanto and GMO technology into maize agriculture, Luna asserts, will not only result in a loss of biodiversity but has already had a profound impact on the health of Indigenous peoples and cultures, specifically in the Oaxaca region of Mexico (Luna 2014b).

Luna problematises the idea of the ecological body as a happy, healthy, white body in the world. However, some of her critique, in her rejection of new technology and new developments in biotech, could be read as techno-phobic or nostalgic for time and labour-intensive traditional growing practices of the past. However, this idea ignores Indigenous peoples' use of technology in farming for centuries including pre-colonisation (Pascoe 2014: 47–48). It is the forces of neoliberal capitalism perpetuated by the US government, which are reasserting a neocolonial agency particularly in regard to the North American Free Trade Agreement (NAFTA), and the loss of biodiversity which are the targets of her piece. As Alaimo (2013) suggests, the ecological body is intertwined with social and political entanglements, which Luna amplifies in the work by revealing the violent effects of globalisation, colonialism and neoliberal market forces on the body that is inseparable from the ecological-material world.

Luna's body is also a gendered body, female-presenting. She represents vulnerable human and more-than-human bodies. As Vandana Shiva asserts, biotechnology, such as Monsanto's seeds and herbicide, is a new form of colonisation, particularly for Indigenous women, and 'these new colonies are, in my view, the interior spaces of the bodies of women, plants, and animals' (1997: 5).[5] This biotech 'revolution' could lead to the instrumental value of women of colour where 'the body parts of women of colour come to be deemed as usable, extractable, tradable, natural resources that the women will be coerced or violently forced into relinquishing for the purposes of those in power' (Stein 2004: 210), recalling histories of oppression and chattel slavery. The risks and effects of these kinds of neoliberal interventions are felt disproportionately on communities of colour. Winona LaDuke interviews Katsi Cook, a Mohawk midwife and activist, on the gendered nature of risk and exposure: 'The fact is that women are the first environment…We accumulate toxic chemicals' (Cook quoted in LaDuke 1999: 18). Biological processes make ciswomen's bodies the first environment as Nancy Langston points out:

> What we do know is that we're all in this together: the atrazine that gets sprayed on my neighbor's cornfields ends up in the river water, then in the fish, then in the herons and the raccoons that eat the fish—and it also ends up in my breasts, my belly and my blood. What's out there in wildlife and wild places is also in our bodies. (Langston 2003: 153)

This passage demonstrates the artifice of the nature/culture divide and the way in which there is no 'out there', as what has been relegated to 'out there' is also 'in here', in our food, our lands and then our bodies. However, this 'we're all in this together' configuration also runs the danger of conflating differences as the extent to which we feel and experience the negative effects of our ecological relationships is often unjustly and unequally distributed along gender, class, race and disability lines: it is intersectional as I previously described.

There is a politics of exposure in Luna's work. Alaimo describes practicing exposure as displacing the anthropocentric idea of the human as the male liberal subject: 'Exposures may be differential, uneven, or incommensurate; yet to practice exposure entails the intuitive sense of the philosophical conviction that the impermeable Western human subject is no longer tenable' (2016: 5). Exposure is acknowledgement and understanding of the material agency of those human and more-than-human actants in and around us, understanding the ways in which we are exposed and what agential forces enable this bodily exposure (social, political, environmental). 'To occupy exposure as insurgent vulnerability is to perform material rather than abstract alliances, and to inhabit a fraught sense of political agency that emerges from the perceived loss of boundaries and sovereignty' (Alaimo 5). Luna enacts material alliances through her performance, use of imagery and her body, between Indigenous peoples and cultures, the plants and land they live off and the global power inequalities that structure their lack of political agency. Luna's vulnerability is a strength of her piece: she understands the intimate interconnection between bodies and is able to dismantle the false binaries that prop up Western capitalist logic.

Heroes and Saints

Heroes and Saints (1992) by Cherríe Moraga enacts parallel themes of exposure of gendered and racialised bodies. The play is set in the fictional town of McLaughlin in the San Joaquin Valley of California, based on the town of McFarland, which became known as a 'cancer cluster'. The United Farm Workers attributed this to pesticide poisoning from working in grape agriculture: 'Within a ten-year period from 1978 to 1988, a highly disproportionate number of children were diagnosed with cancer and were born with birth defects' (Moraga 1994: 89). This resulted in the United Farm Workers grape boycott of 1988. The exposure of Chicano/a farm workers to pesticides and the relationship between poisoned bodies and poisoned land, is the subject of Moraga's play, which was first performed in 1992 at El Teatro Misión of San Francisco.

Heroes and Saints' protagonist is Cerezita Valle, described as 'the head', a teenager who was born without arms or legs as her mother, Dolores, was a farm worker exposed to pesticides when Cerezita was in the womb. May describes her as 'material evidence and reminder of the impact of toxins on the health and well-being of workers and their families as well as the land, but she is also a metaphor for a society that denies the body' (2007: 101). The farm workers' exposure to pesticides stemmed from both their work (the only means of earning a living in the area) and their homes, government-subsidised housing built on a waste site of industrial agriculture with contaminated tap water. In the play, to incite action on their exposure, the community, or 'el pueblo', create horrifying images by suspending the bodies of the dead children on crucifixes in the fields, signifying the connection between land and bodies including the land and death. The Chicano/a workers are subjected to environmental racism through the systemic forces of the low-flying crop dusters spraying pesticides overhead that punctuate the action and the law enforcement that monitor, beat and shoot the protestors, protecting the interests not of the land or el pueblo, but of the grape growers, 'the bosses'. Cerezita's brother Mario has AIDS, which was spread (or neglected to be controlled and treated) through the same systemic forces that oppress the worker. Whiteness, wealth, power, straightness, able-bodiedness and land 'ownership' are the qualities that merit protection, at the expense of all those bodies (human and more-than-human) that do not possess them. Moraga makes an explicit connection in her prose and poetry to land as home and home as land:

> Land remains the common ground for all radical action. But land is more than rocks and trees… For immigrants and native alike, land is also the factories where we work, the water our children drink, and the housing project where we live. For women, lesbians, and gay men, land is that physical mass called our bodies. (1993: 173).

The play is doing what Alaimo calls 'performing exposure as an ethical and political act' (Alaimo 2016: 5). In the last moments, Cerezita, previously sheltered and trapped at home, speaks to the people and goes to hang the dead body of her niece on a cross in the field amidst machine gun fire from law enforcement in helicopters. She is politicised through the play by witnessing and experiencing the material relationships of her place. The pesticides caused her disability and the death of others, making the material alliances deathly. The ethics of the exposure to toxic pesticides by those vulnerable and/or marginalised are questioned in the play through authorities' support of the 'the bosses' rather than workers. There is no public outrage over the deaths of the children outside the community. Cerezita's lack of a body represents the denial of subjective embodiment: 'to dramatise oneself in place in this way is to critique the rational, disembodied Western subject's presumptions of mastery or at least objectivity that is, supposedly, granted by detachment from the world' (Alaimo 2016: 5). The ability to presume a detached stance from the more-than-human world, from the land, is a privilege that the Chinano/a farm workers do not have access to.

The body as home/ecology is, as Di Chiro suggests, an 'apt metaphor and material ground' for an 'embodied ecological politics that articulates the concepts of diversity, interdependence, social justice, and ecological integrity' (Di Chiro 2010: 200). Home is one way of translating *oikos*, the root of 'eco', as I discuss in the following chapter. 'The body can be reclaimed and refigured as home—that desired place of connectedness, family, and wellbeing—with full realization that the body/home is sometimes the site of exposure to just the opposite: abuse, hunger, polluted water and air' (Di Chiro 200). Eli Clare also warns against a cosy, safe, ableist understanding of the body as home by drawing attention to the daily ways in which the body as home is not safe, healthy or controlled: 'The body as home, but only if it is understood that bodies can be stolen, fed lies and poison, torn away from us. They rise up around me—bodies stolen by hunger, war, breast cancer, AIDS, rape; the daily grind of the factory, sweatshop, cannery, sawmill; the lynching rope; the freezing streets; the nursing home and prison' (1999:

12). Clare goes on to write that stolen bodies can be reclaimed, memorialised, used to strengthen our will to transform deadly and dangerous stereotypes lodged in our bodies. *Heroes and Saints* demonstrates an embodied ecological politics that does not place the body as a cosy, safe home but as a site of violence and resistance to systemic racial and ecological oppressions, with hope for transformation.

Performing exposure in both Luna's *NK603: Action for Performer & e-Maiz* and Moraga's *Heroes and Saints* is a radical way of revealing and resisting the material agency of pesticides, GM seeds and the political and social forces that distribute and protect them. Dramaturgically, these two works generate meaning in different ways. As multimedia performance art and as a scripted, linear dramatic play both illuminate the way bodies are part of complex ecologies that include multiple contexts which produce exposure and vulnerability in different forms. Humans are not in a hierarchical great chain of being to 'master' the more-than-human world; rather, exposure operates across and between bodies, intercorporeally. For Weiss, 'to describe embodiment as intercorporeality is to emphasize that the experience of being embodied is never a private affair, but it is always already mediated by our continual interactions with other humans and nonhuman bodies' (Weiss 1998: 5). We are exposed to the continual interactions with bodies of all forms which co-constitute our bodies in complex and multifarious ways.

Still Dance with Anna Halprin

A fallen log on the forest floor, covered in pine bows, twigs and other forest debris, looks innocuous at first glance, but another look reveals a face and body lying on their side. It is Anna Halprin (in her eighties at the time) lying in stillness. Her nude body is painted and covered with the material of the forest floor, allowing her body to blend into place, becoming part of the forest. This is Eeo Stubblefield's photograph, *Fern Series #8* (1999), part of her *Still Dance* series with Halprin. She is exposed to the elements of the forest floor and the camera taking her photograph. This work draws on an understanding of the body as embedded in site, reaching out for primordial connection. However, in doing so it presents romanticised bodies and environment: the image of a white, able-bodied woman with a normative body can invoke an exclusive, essentialist and reductive idea of which bodies are 'natural' in relation to ecology and nature settings.

An American dancer and choreographer, Halprin's career in dance theatre and applied dance has spanned more than 40 years. Halprin has long been interested in how dance can mediate our relationship to the environment. Halprin's work with her Sea Ranch Collective in California provided a space to create work that responded to relationships with place and the more-than-human world. She also sees the environment as an active partner in her site-based performance work, 'a complex combination of relationships' (Halprin quoted in Poynor 2009: 126), one that provides spontaneity as the living world is always in flux, asking the performer to respond instantaneously rather than attempting to control the natural elements. At the core of the work is the premise of the interconnectedness of human beings and the living world, reflecting Halprin's non-anthropocentric belief in recognising humans as part of a larger whole rather than at the centre.

Still Dance with Anna Halprin (1997–2000) is a collaboration with performance artist and photographer Eeo Stubblefield (and based on her original work) in which Halprin immerses herself in a number of locations, blending into them. Wearing costumes and body paint that echo her surroundings, she lies in a fallen redwood tree covered in moss and forest debris; on a beach getting tossed with the waves, wrapped in white muslin with bent limbs, in the Driftwood Series; beside a stream; on a hilltop set against the sky dressed in a white flowing fabric reminiscent of clouds. There are few conventions of dance performance as there is no choreography or timed movement, no music or set rhythm, no designed lighting. Stubblefield understands it as:

> a conversation with the land… As the performer you open to its power articulating through your body its ground, its water, heat, wind, color, sound, season, the time of day… And it is not a process of forcing the land to serve as a metaphor for an internal human state; nor is it using the land as a backdrop for an abstract set of aesthetic principles. It is rather, a collaboration between the inner and the outer worlds. (n.d.)

It is a piece that exists in two forms: the improvised dance between the dancer and the more-than-human world of a specific place and the visual form of the performance captured through Stubblefield's photography (the primary way it reaches an audience).

The piece enacts the duality of stillness and dance—seemingly a paradox—but it is the juxtaposition that creates the affective image. Arden

Thomas (2012) argues that it is through the still, improvisational dances that a radical, political act is embodied in relation to the more-than-human world. For her, stillness resists the capitalist imperative of expeditious production and consumption while also gesturing towards a slowed down temporality of the more-than-human world, between the human and nonhuman, revealing the reciprocity in their relation. 'The *Still Dance* in nature resists and defies capitalism's restlessness as well as its voracious desire to consume and reproduce' (Thomas 2012: 118). The neoliberal body is valued for labour and consumption practices, or how much it can produce and consume. *Still Dance* reframes this as a living, sensing and perceiving 'body subject' (after Merleau-Ponty) that is more than their labour or consumption and resists the body/mind dualism, particularly evident in an elderly, aging body that can no longer perform as the 'productive' labour-producing subject. Thomas also sees the piece, after Val Plumwood's ecofeminism, as rejecting the binaries of oppressive ecological subjugation which 'denies a dependency or interconnection between self and other, human and nature' (2012: 118). For Thomas, this work resists 'modernity's crippling speed' through slowness and stillness where 'the body finds its being, its becoming, its living, not in modernity's kinetic excess or capitalism's hyperactive consumption of nature, but in the still spaces between the human and nonhuman' (119). The stillness becomes a way of interacting with the more-than-human world outside of neoliberal imperatives. The slow, still dance where Halprin's body moves or not, as the waves, as the clouds, as the dead tree, counters the productivity imperative that shapes the temporality of neoliberalism. It also reframes 'nature' as a backdrop for consumption or a resource for extraction. In this sense, stillness is a resistant act that reframes the reductive dualism of human/nature. The photographs of the piece are enthralling and make me want to shake off the oppressive forces of capitalism and find the nearest forest to immerse myself in. The sense of liberation would likely only be fleeting though because ecological relationships are inextricable from sociopolitical contexts, including capitalism.

Halprin and Stubblefield seem to be representing the more-than-human world while also embodying it. Acknowledging the relationship between the living world and living body, they are enacting Abram's concept of 'becoming earth' by 'tuning our animal senses to the sensible terrain: blending our skin with the rain-rippled surface of rivers' (2011: 1). The blending of Halprin with the more-than-human world in and around her, through the use of theatrical (or mimetic) conventions, utilises the space of performance to reveal the assertion that we are part of the living world through trans-corporeality. There is the suggestion of more-than-human agency, as the trees, mud, clouds and ocean leave impressions on Halprin's body and structure her movements. By taking on the characteristics of the non-humans in a specific place, she is enabling a form of dis-anthropocentricism, or horizontal relationships between her and the actants in place. However, this work is open to the criticism of romanticising nature. The still photographs that document the work tend towards valorising the beauty of 'natural landscape': a secluded beach, a dense redwood forest, a streaming waterfall, a vivid blue sky, and an able-bodied white woman's connection to it and the female association with nature. The images of *Still Dance* are bright and compelling. I am attracted to this version of 'nature' as secluded, dense and 'wild', yet I know it is a reductive and limited version that can be essentialist and exclusionary. The chosen landscapes depict an Edenic version of 'nature', rather than one that includes humans they seem to suggest that humans should blend into this romantic image. This Edenic image is one of nature as a balanced circuit, which can be restored to a balanced state of harmony and is valued for its beauty. There is no stability and stasis in 'nature'; it is in a constant state of change. What would it mean to 'blend' into a concrete traffic island? Or a sewer grate? Or a housing project? In a sense then, while *Still Dance* is ecological in its resistance to capitalist grand narratives of value and productivity, it romanticises an idealised version of 'nature', which may be quite distant to the ecological material reality and only accessible to a privileged few.

There is also an element of exposure in the piece. Alaimo writes about naked eco-protests (often naked bodies spelling out words in a field or beach, such as 'NO GM') that 'extend human corporeality into actual places, practicing the art of exposure, dramatizing the allied, coextensive permeabilities of human/animal/environment' (2016: 67), which could also be read in *Still Dance*. However, the naked bodies in these protest works are often gendered and racialised. The naked bodies in 'nature' lean towards the 'trope of the body as wholesome and natural—a trope saturated with straight, white privilege' (Alaimo 2016: 71). This is evident in *Still Dance*, not only is the environment romanticised as a pure space of beauty, the body is also romanticised as 'wholesome and natural', framed in a way that is not 'excessive' or objectifying but incidental. Alaimo goes on to suggest that 'perhaps these scenes of exposed but not "excessive" flesh, made up of predominantly white bodies arranged across landscapes, flatten racial categorization and disrupt codes of visuality by presenting nothing to see' (Alaimo 2016: 72). Carolyn Finney argues that normalising the white body in 'nature' through images in popular culture and culture at large (the exclusion of people of colour from the Green Issue of *Vanity Fair*, and nature magazines such *National Geographic*, *Backpacker* and *Outside* for example), shape ideas of who 'belongs' there and who does not, often resulting in racialised environmentalism that excludes non-white bodies (Finney 2014: 1–2). An intersectional understanding of bodies in environments is needed to consider how these works are making meaning.

Still Dance has clear parallels to Cuban-American artist Ana Mendieta's earth body works in which she imprinted her body on different landscapes in her *Siluetas Series*. She included her own body as part of the landscape and used the traces of it to create sculpture and performance works captured through photography and film. Mendieta's work predates *Still Dance* with over one hundred *siluetas* made in Iowa and Mexico between 1973 and 1980. These include carvings of her figure into stone or clay landscapes, lying nude face down in a creek, raised mounts of her form on hillsides and shores, setting fire to her imprint and lying nude at the base of a tomb at Yagul archaeological site with her body covered in white flowers that look as if they are growing out of her. In *Arbol de la Vida*, or *Tree of Life* (1976), she plastered her body in mud and stood with raised arms at the base of a large tree facing the lens, blending into the trunk. She writes about this as a kind of return: 'I have thrown myself into the very elements that produced me, using the earth as my canvas and my soul as my tools' (Mendieta and Clearwater 1993: 41). These works draw on similar themes to *Still Dance* by examining the relationship between femininity and 'nature', although Mendieta was also interested in the imagery of women in Indigenous Mesoamerican cultures, choosing sites and symbols to reflect this such as the tomb at Yagul archaeological site in Oaxaca, Mexico and the iconography of the female form from Afro-Cuban imagery. While her contemporaries (such as Robert Smithson, Michael Heizer, Walter De Maria) made large-scale earth works, Mendieta's body and identity was her focus. While her work is read in multiple ways by different people, for some, such as Muñoz, 'it is all about the accounts of her small brown female body manifesting itself in a field of possibility dominated by often hostile white men' (Muñoz 2011: 191). Mendieta's series is open to the same essentialist critique as *Still Dance*, as she subscribed to an 'idea of feminized nature', which Susan Best argues 'allows her to posit alternatives to patriarchal culture in the name of the feminine, that include a kind of ecological sensibility that emphasizes the reciprocity between body and land, a resistance to colonialist conceptions of land and territory' (2007: 58). Themes of loss, displacement and erasure permeate the work and help to complicate the image of a white woman immersed in 'nature'.

British artist Ingrid Pollard's landscape photography, such as *Pastoral Interlude* (1987), adds a further intersectional understanding to art works of bodies in 'nature'. Her work photographs Black bodies in picturesque English country landscapes, commenting on race, representation and national identity. *Pastoral Interlude* is a hand-tinted photograph that evokes a quaint heritage of the romanticism of the English countryside and who 'belongs' there. In the image, a Black woman is sitting on a stone wall, holding a camera on her lap, dressing in walking gear, gazing out to the left of the frame. Behind her is a wire fence with rolling green hills beyond. The photograph is accompanied by a caption: 'It's as if the Black experience is only lived within an urban environment. I thought I liked the Lake District, where I wandered lonely as a Black face in a sea of white. A visit to the countryside is always accompanied by a feeling of unease, dread' (Pollard 1987). Her body is exposed in these landscapes of quintessential 'nature' as it does not fit into the dominant image of who gets to occupy and enjoy these spaces.

Pollard's work illuminates the way in which stereotypes have excluded Black Britons and Americans from participating in the 'Great Outdoors' as well as environmental movements, constructing them as White spaces and issues (Finney 2014). Finney argues that 'These negative stereotypes, often grounded in racist practices from our past, have bled into present-day narratives (e.g., Hurricane Katrina) perpetuating old grievances and creating new roadblocks to a more expansive consideration of African American contributions to our ongoing environmental challenges' (2014: 11). In the late nineteenth century, with rapid changes such as 'immigration, urban expansion, industrialisation and women's increasing economic independence...outdoor pursuits came to serve as a new space for elite enactments of white male superiority', with activities such as 'hunting, fishing, climbing' (Mortimer-Sandilands and Erickson 2010: 3). The conjoining of these types of outdoor activities with a white, male superiority is discursively and representationally constructed. The predominant images and assumptions in the US (and I would argue other places in the Global North including Canada, UK, Europe, Australia) about ecology and environment mean that 'we have collectively come to understand/see/

envision the environmental debate as shaped and inhabited primarily by white people' (Finney 2014: xii). Finney contends that not only are these assumptions and perceptions exclusionary and render people of colour's participation in environmental issues invisible, they are foundational to 'how we define the "environment" and how we think of ourselves in relationship with the environment' (xii). These configurations get cemented in the cultural imaginary through popular media and representations as the norm. Whiteness as a perspective becomes the way of understanding and interacting with the 'environment'. Cultural representations can also critique, question and deconstruct stereotypes and historical exclusion as Pollard's work does. Ecodramaturgies can act as 'sites of informal knowledge production' (Finney 2014) which can reveal how these forces shape contemporary ecological relationships.

Although Halprin's ecosomatic work is alluring in its beauty, it belies a privileged perspective and bodily experience that is in danger of reinforcing harmful stereotypes about which bodies are 'wholesome' and 'natural' and belong in these places of natural beauty. Read with Mendieta's *Siluetas Series* and Pollard's photography, it is clear the ecological body is coded and shaped by the world around us, ecologically and by gender, race, ableism, neoliberalism and other social systems of 'naturecultures' (Haraway 2003). Different bodies are exposed in these systems, leaving grapes, corn, Mayans, *campesinos*, Chicano/a workers, Black British walkers and dancers vulnerable in different ways.

Participation and Immersion

We are always already immersed in the more-than-human world. We cannot separate ourselves from this immersion. As bodies, human and non-human, we are exposed and made vulnerable through this immersion which is differentiated and unequal. The bodily immersion in the ecological world is influenced by socio-political structures that determine which bodies are exposed and valued.

This immersive embodied engagement is characterised by participation in the more-than-human world. Through the simple act of breathing we are always participating in the circulation of oxygen and carbon dioxide in the atmosphere. Abram (1997) identifies participation as the defining quality of our sensuous immersion with the world—participation both in our sensible engagement with the more-than-human world, as well as there being a participatory quality to the interplay of the senses themselves. Drawing on participation as used by early French anthropologist Lucien Levy-Bruhl, Abram describes the more-than-human extension to participation: 'particular plants, particular animals, particular places and persons and powers may all be felt to participate in one another's existence, influencing each other and being influenced in turn' (1997: 57). We all participate in each other's being-in-the-world in differentiated ways, and when we consider this in relation to the ecology of the more-than-human world we see that we profoundly influence, and are influenced by, our ecological relationships. From the air we breathe to the water we drink to the sunlight on our skin to how we vote, spend money, travel and eat, we are affected and affective in destructive and pleasurable ways. Ecodramaturgies can mediate our understanding of the way in which we are immersed and entangled with more-than-humans in an ecological sense and the way in which our perspectives on that immersion is shaped by social and cultural formations and representations. The performances in this section, *Wolf* (2013) and *World Climate Change Conference* (2014), offer different ways of uncovering the often invisible ecological relationships we are immersed in, through ecodramaturgical strategies of site-specificity, humour, performance and storytelling. They also facilitated different forms of participation and brought to light socio-political contexts and structures that can enable and disable participation.

Wolf

In September 2013, an audience of 108 met outside Krannert Art Museum at the University of Illinois at Urbana-Champaign (US). Two large coaches were awaiting the audience. As they entered they were each given a devotional wolf medallion: a circle of wood with a howling wolf etched on it attached to a piece of string to be worn around the neck. Khakied performers dressed as park rangers greeted the audience and narrated the 40-minute drive to the Allerton International Biosphere. Along the drive, the small video screens mounted to the roof of the bus played videos of wolf mythology, a wolf expert explaining their behaviour, a farmer who lost cattle to wolves, the lyrics to a song and a chart of wolf facial expressions. The rangers told stories and gave background information about Allerton International Biosphere and the wolves let loose there. Although they speak with authority and knowledge, something seems slightly unreal about the stories they are recounting. The audience is asked to sing along to *Home on the Range*, with a darker second verse. They are asked to howl like a wolf and practice wolf facial expressions of dominance and submission with each other. Arriving at the woods of Allerton Biosphere, the performance increases in its mysticism and playfulness. The audience is divided into herds: moose, bison, deer and elk. They undertake a group walk through the woods in their herds, eventually heading to a large barn at the centre of the wooded area for an hour performance featuring the story of Loki and Fenris; a golden retriever who killed the wolf population on Isle Royale on Lake Superior; and different incarnations of the Big Bad Wolf. It only gets more surreal from there.

Wolf is part of Deke Weaver's *Unreliable Bestiary* project, which aims to make a performance for an endangered animal or habitat for each letter of the alphabet. Weaver, an American theatre and media artist, describes it as a lifelong project. It is inspired by the chilling prediction that half of all species will be extinct by 2050. In addition to the site-specific live performances, there is a growing encyclopaedic set of films, books and solo-performances that make up the project. Since 2009, *Monkey, Elephant, Wolf* and *Bear* have been performed and *Tiger* is in development. The solo performances have toured nationally and internationally, part performance and part documentation (often with film footage and slides).

Weaver positions his work as an ecology project, not as representing animals as metaphors for human conditions. He positions it as a gesture 'presenting a tiny sliver of our current catastrophic loss of habitat and biodiversity' under the assumption that climate catastrophe will mean the eventual loss of more than half of the world's species (Weaver n.d.). For him 'animals and their stories are embedded in our environmental, economic, political, and judicial systems, the systems thoroughly enmeshed with each other' (Weaver n.d.). Weaver's approach, however, is novel. 'I'm trying to find ways to charm people into realizing the complexity and urgency of our situation. I want the project to turn people on instead of shutting them down with fear. The *Unreliable Bestiary* is using humor, poetics, and plain-old wonder to inspire people to live differently' (Weaver quoted in Rothfels 2014: 182). The charm and humour of Weaver's project is an effective strategy for revealing the layers of complexity in current ecological contexts, as well as the interconnections between bodies such as elephants and humans, wolves and stars, monkeys and deities.

Wolf was an approximately four-hour performance experience in three locations: bus, woods and barn. Performers played the Rangers of Allerton International Biosphere, including Weaver. Dressed in khakis complete with official-looking embroidered patches on their arms, the ranger's performance style was convincingly sincere—for the most part. They began by describing the geophysical history of the area 15,000 years ago, when the glaciers left the area to form the Great Lakes. Allerton International Biosphere is known as 'the zone', comprising 1500 wooded acres and one of the seven wonders of Illinois, 'a little bit of art, a little bit of nature' the audience is told. As the buses drive through Illinois, the audience is told '80% of land is used for farming, that's 28 million acres of land' with very little 'wild' areas (Weaver 2013). 'We are about to sail across a sea of corn, an ocean of soy, giant patches of genetically-modified monoculture' says Weaver. He brings up the farming industry of the area in a way that is sympathetic to farmers, 'if I could figure out a way for my land to give me more gain I'd probably do it' (2013), allowing for multiple perspectives to be put into dialogue and calling attention to the interconnected ecological relationships of the area. The audience is invited to sing *Home on the Range* 'in memory of the people and animals who once called this area home' (2013). The song begins with the lyrics on the roof mounted video screens of the bus. The audience sings along, although slightly hesitant with the second verse, a darker rewrite of the original about loneliness and isolation.

Questions of representation arise in the way wolves are perceived and valued (or not). 'The real animal often doesn't have much to do with the animal in our stories or movies. Unfortunately, it's usually the myth that determines how we treat the real animal' (Weaver 2013). The stories and representations put into dialogue in the performance demonstrate the myriad of ways wolves are fundamental to mythology of all kinds. As wolves live in all different lands including prairies, tundra, and the Mediterranean, they feature in many cultural imaginaries: 'Wolves exist in the stories and cultures and myths of human settlements all over the Northern Hemisphere because wolf territories exist pretty much anywhere you can imagine' (2013). Wolves are part of many mythologies and are represented in Western popular culture as villains, killers, pests and as a metaphor for 'nature'. The performance of *Wolf* deconstructs these representations, revealing their nuances and influences through humour and storytelling (Fig. 4.2).

The Big Bad Wolf features in many of the stories told in the barn performance, in many different incarnations, but never as the one-dimensional killer represented in fairy tales. For example, in one noir story, told from the wolf's perspective, he meets a hunter at the bar, furious at the years of violence and murder of his kind but acting cool on the outside. As Weaver

Fig. 4.2 WOLF dancers (from left) Jennifer Allen, Laura Chiaramonte, Angie Pittman, and Jessica Cornish in *Wolf* by Deke Weaver, Allerton, Illinois, 2013. (Photo: Valerie Oliveiro, courtesy of Deke Weaver)

as a ranger points out on the bus, wolves hunting their food is not necessarily violent, despite how they are portrayed. He links their actions to human behaviour:

> you and I go to the fridge, pop something on the stove for dinner. A wolf running 30 miles per hour pulls down a 900 lbs elk with their teeth. That's what they do, they don't need to get angry about it. That'd be the equivalent of you screaming to your pork chop sandwich and stabbing it a dozen times before you eat it. You don't need to do that. (2013)

Weaver draws attention to the cultural representation of wolves as violent predators, who are almost always the villain. Through playfulness and identification, Weaver undermines this idea by telling stories in which wolves are victims, innocent links in the food chain, and creators of the universe. *Wolf* prompts the audience to rethink the place of wolves within their cultural imaginaries, in order to question their endangered status. In 2018, there was an estimated 35 red wolves left in the wild in North

Carolina with calls from a small group of landowners to remove protections of these 'nuisance animals' to be shot and restrict their territory (Drew 2018). In Cary Wolfe's response to Weaver's *Monkey* (2009), he quotes Cora Diamond on the 'difficulty of reality' (Wolfe 2014: 159). For Diamond, this is 'the awareness we each have of being a living body' and this body is exposed and 'carries with it exposure to the bodily sense of vulnerability to death, sheer animal vulnerability, the vulnerability we share with them' (Diamond 2008: 74). Acknowledging this shared vulnerability and exposure with other animals, in light of how we treat them, is 'capable of panicking us' (Diamond 74). This shared sense of vulnerability is also present in *Wolf*, beginning with the stories of the misrepresentation of wolves as 'beasts of waste and desolation' (Weaver quoting Theodore Roosevelt 2013), and then by putting the audience in the shoes of wolf prey, and finally by the first-person stories performed in the barn with Weaver as multiple iterations of a wolf. Humans playing the wolves at the Allerton Biosphere perform idea of the shared bodily sense of exposure.

Weaver puts the politics of wolves in dialogue. The stories of wolves being hunted and eradicated point to the idea that violence towards wolves is based on the perception that they are 'bad'—as this is how they are largely represented in popular stories and cultural media. Weaver notes to the audience 'depending on how you grew up, you might feel differently about wolves. Sometimes those opinions get so strong that the wolf is pretty much wiped off the Earth' (Weaver 2013), before recounting the story of wolf extinction in Yellowstone National Park. In the 1920s the policy was to shoot wolves, 'extermination seemed like a good idea to white folks' (Weaver 2013), with the last one being shot in 1926. Then things began to change in the park as the elk population grew and changed the landscape by eating more of the vegetation, leaving no young trees. This then affected the rivers, with no tree shade or tree roots to hold the banks, the rivers changed flow, as well as fish and microbes. Wolves are a keystone species that maintains ecological diversity in a biosphere. After being listed as endangered in the US in the 1980s, they were reintroduced to Yellowstone in 1995. This story is directly in dialogue with a video interview with farmer—also played by Weaver—who talks about the overpopulation of wolves that kill pets and cattle. The farmer fears for child safety, saying when that happens, something will finally be done about the wolf problem. Weaver does not vilify the farmer, saying that management of the wolves is not as 'black and white as you might think', it is complicated. He would be upset too if wolves killed his cattle. He also acknowledges the capitalist influence on representations of wolves. Invoking the Disney-version of the Big Bad Wolf as a simplistic villain, in the barn, Weaver tells a story of Snow White wanting to get rid of the Big Bad Wolf in order to protect her investment, as it is neoliberal capitalism that governs the Disney forest. Reimagining the familiar trope of the Big Bad Wolf is also a decolonising gesture as Weaver recuperates the wolf as an important figure in creation mythology (in Greek, Pawnee and Nordic people's mythology) and ecosystem diversity, rather than a pest who kills livestock.

The immersive performance event creates space for dialogue and response. The audience is asked to participate in a number of different ways: be wolves, interact with wolves, and be prey for wolves. The audience is told when they enter 'the zone' of Allerton Biosphere, the energy force is so strong they might experience euphoria, mood change and weightlessness. They are told they will be given a chance to feel like they are being hunted. 'There is nothing that makes you feel more alive than realising you're part of the food chain' (Weaver 2013). The audience is told to get their senses on high alert and walk silently through the forest. This playfulness belies an ecological reality: humans are part of the food chain. Removal from hunting other animals for food does not release humans from the 'shared vulnerability' of being part of a complex ecological system in which we have some shared exposure but also disproportionate power.

As the audience herds walk silently along one of the wooded trails they are stopped by the ranger with a wolf sighting. A human actor in a furry wolf costume is lying across the path. The audience is told the wolf is playing dead as it has succumbed to a trap set to study the wolves. The rangers tell the audience to be careful as the wolf's fur is full of a high concentration of 'the energy', and if ingested, they might achieve transcendence prematurely. The wolf actor then gets up and runs quickly away on two legs, sporting a pair of trainers. I laughed aloud at the oddness and unexpectedness of this moment. The humour is created by the lack of trying to be wolf-like by the wolf-performer and the seriousness in which the ranger approaches the 'wolf'. The humour draws on the uncanniness of the wolf–human hybrid and the audience role of observers and visitors to The Zone. The walk through the woods continues with other 'wolf' sightings.

This novel, bodily immersion and participation in *Wolf* is a type of somatic knowledge, as Una Chaudhuri describes in Weaver's piece *Polar Bear God:* 'The pared-down style of the piece highlights the potential for performance to offer a kind of somatic knowledge, a way of understanding the Other by going beyond rationalisations and abstractions to embodiment and physicalisation' (2012: 52). The performers playing wolves, the human–wolf hybrids, embody wolves without parody or abstraction. The audience could also access somatic knowledge through the invitation of the performance. At one point along the walk in the forest, the audience is told to 'feel the zone's energy through your feet or put your ear to the ground' (Weaver 2013), which many did. This quiet moment of group reflection was a novel way to experience the space in the semi-fictional performance. It also had ritualistic qualities which were both playful and respectful. The physicalisation and embodiment of wolves was also present in Simon Whitehead's performance piece *Louphole* (2010, 2013), which used a 'public howl' to howl like wolves to reflect on the extinction of wolves from the Welsh landscape. The invitation to the public to howl with Whitehead at dusk, 'examined ways in which the memory of this animal may still reside in embodied or folk idiom and how participatory performance and voice may evoke recollection' (Whitehead 2018: 120). In Weaver's *Wolf*, rangers brought each herd to a central meeting place using horns and a series of noise makers. The group then proceeded to the barn at the centre of The Zone, where a pack of wolves (human actors) were playing and running outside. Inside the barn there was a wolf shrine with models and maps of wolf territory and a golden wolf statue. In the performance space, a circle was outlined on the wooden floor with a wolf-actor lying in the middle. Some small animal figures were placed around the circle and images of the solar system played across two screens set in traverse.

Through the imaginative mix of fiction and reality, Chaudhuri writes that 'animal acts convey new knowledges through new bodily experiences in space and time. They invite us to explore new habitats, where we might practice more imaginative and ethnically ways of life' (Chaudhuri 2014: 10–11). The performance invites the audience to experience a new habitat and wolf perspective (and the stories that render it) in relation to their own place in the ecosystem and the power dynamics at play. Weaver's stated aim for the project is to 'inspire people to live differently', perhaps acknowledging a non-anthropocentric perspective that takes into account the other species in and among us and might counter the climate crisis and its drive to extinction.

The Disney stories of the wolf as 'big and bad' makes it easier to deny the 'shared vulnerability' of wolves as fellow animals by hunting and exterminating them as a threat to capitalist imperatives. The performance experience of *Wolf* tells other stories of wolves, exploring wolf as symbol and material relationship. The barn performance ends with the line 'everything is connected' set against a video of the solar system. Instead of a simplistic or glib refrain, 'everything is connected' becomes a linking thread for the piece, connecting the three parts and experiences of park ranchers on a bus, hiking through a forest looking for wolves and then being in the barn and witnessing multiple takes on wolf stories. It also reveals the place wolves have in mythmaking and the world-building, connecting Fenris the wolf to starwolves and the solar system. In the shadow of climate change, these connections are complex, humorous, scary, surprising and sometimes absurd. *Wolf* encompasses and expands on ecological relationships, allowing the audience to situate themselves in the messy tangle of interwoven narratives and materials. For Chaudhuri, 'the underlying principle of *The Unreliable Bestiary* is that live, embodied performance can reveal and perhaps revise the ideological mechanisms that have allowed human beings to bring so many other species to the brink of extinction' (Chaudhuri 2016: 184). Playing with ecological relationships through immersive and experiential ecodramaturgical structures allows for expanded ideas of humans and animals, blurring the established binaries that rest on ideological foundations of extractivist human hubris and hierarchy.

World Climate Change Conference

The United Nations have held an annual climate change conference since 1995. For most of us though, world climate change conferences seem far removed from our day-to-day lives. We might hear that they are happening via a headline or news article or perhaps follow the agreements coming out of the bigger gatherings (such as the Paris Agreement) but generally they are not an experience most people are intimately familiar with. Their processes seem abstract as they often appear to be comprised of vague diplomatic negotiations. However, they do have the ability to set the global agenda on government responses to climate change. What would it be like to be a participant in one, embodying climate change figures and decision makers? German audiences had a chance to find out in Rimini Protokoll's *Welt-Klimakonferenz* (or *World Climate Change Conference)*, which facilitated audiences to think about their participation in climate change and global politics. Taking a different form of ecodramaturgies and participatory performance, *World Climate Change Conference* (2014) allowed audience members to enact the global power relations of climate change negotiations.[6] Rimini Protokoll are a Berlin-based arts ensemble who are interested in making new forms of contemporary performance. Disheartened with the lack of collective agreement and action on climate change, they staged their own world climate change conference in Hamburg, Germany in 2014 and 2015 ahead of the United Nations Climate Change Conference, COP21, which took place in Paris in December 2015. Over a 3-hour period, more than 700 audience members per performance were briefed by experts, experienced climate change modelled scenarios and given a chance to negotiate emissions reductions. Upon entering the theatre, audience members were invited to choose a delegation out of 196 nations, which they would represent for the whole of the performance, negotiating on their behalf. For example, audience-delegates could choose to:

- represent India, persisting with efforts to undermine the plan of action by introducing feeble formulations with minimum commitment…
- Canada or Japan, who have long since turned their backs on the Kyoto Protocol…
- in Paris you could also champion an issue on behalf of Bangladesh, one absolutely not to the industrial nations' liking: compensation for the environmental damage caused by storms and floods. (Rimini Protokoll 2015)

An opening ceremony at Deutsches SchauSpielHaus Hamburg theatre involved a panel of scientists who sat behind the white table, adorned with the UN climate conference logo, each with a microphone, name plate and bottle of water, mirroring the visuals and semiotics UN conferences. Information was projected behind them, all in the UN colours of white and pale blue. Audience members sat in the theatre in their delegations marked by their countries' flag. They were given an information booklet about their nation on a red lanyard with detailed information about land, climate, population, industries, standard of living and greenhouse gas emissions (with global comparison). They were told in the opening ceremony by Dr. Florian Rauser, physicist, that 'our challenge here today is to find a solution' but 'it's all rather complicated. It's not just a question of turning off greenhouse gases—the issue is related to virtually everything'

(Rimini Protokoll 2015). In addition to committing to CO_2 emissions reduction, the delegates were also told that if their nation could not reduce emissions to the needed levels, they could contribute to the Green Climate Fund as 'money lubricates every climate conference' (2015).

After the opening ceremony, the delegates were divided up into regional groups and ushered to different briefings in different locations. In a black box theatre space, Dr. Bernd Hezel, physicist, explained to the Western European delegation what changes a temperature increase would bring for them, declaring Norway the 'winner' of climate change because the temperature rise would open up new industries such as tourism and agriculture for them. The Western European delegation sat on a seating bank, looking at the African delegation experiencing the 'Example from Africa'. The African delegation laid down on folding cots wearing headphones, in a circular formation on a revolving platform. Bright floodlights shone from above, simulating the rise in temperature. From a raised platform in the centre of the cots, Kenneth Gbandi, president of the Nigerian Diaspora, spoke to them about the changes in the weather, while also making weather to demonstrate those changes. Clouds followed by heavy rain, made through a fog machine and water from spray bottles were used to weather the delegates. Another delegation heard from Dr. Schirin Fathi, expert on the Middle East, while on a coach travelling between venues. She tells them 'colonialism had an effect on most nations, and resulted in weak social structures' (Rimini Protokoll 2015). She states that the next war will be fought over water as drought increases. This emphasis on understanding the complexity of climate change in social, political and ecological terms was partly achieved through this kind of intersectional ecological thinking.

A panel of scientists was present to establish the context of the conference (including a brief history of climate conferences to date). There were scheduled talks for delegates based on geographic regions as well as lectures or modelled scenarios on environmental conditions such as drought and polar ice melting. This assisted the delegates in gaining an experiential and embodied understanding of the effects of climate change. Through the embodied conference dramaturgy, Vicky Angelaki writes that '*Welt-Klimakonferenz* did not take a symbolic approach to the universality of the issue but rather a corporeal and literal one, with bodies at work on the ground: from discussions about to simulations of climate change conditions, this was a densely lived rather than merely live event' (Angelaki 2019: 43). In the Drought Scenario, led by Dr. Vera Schemann, mathematician, delegates lay down on the folding cots from the 'Example from Africa' while a future scenario was described to them in the first person. 'Everyone here today comes from a country with a water shortage...You have less than 1000 m^3 of water per capita per year' (Rimini Protokoll 2015). A fog machine was used to create clouds to show how drought is caused and affects the whole atmospheric system. Rather than imagining drought as an abstract 'over there' concept, they were asked to imagine the personal: 'You have little water. What would you do?' This, coupled with the experience of weather and atmospheric changes, meant that these abstract scenarios were experienced bodily, physically and personally.

Delegates were then given time to discuss in bilateral meetings to learn from other delegations before committing to their own reductions. When they were finished negotiating, they were asked to sign an *absichtserklärung* (memorandum of understanding) stating the percentages of emissions reduction they were committing to in 2020, in 2050, and the amount of money they would give to the Green Climate Fund. At a plenary session back at the theatre, Rosemarie Benndorf, an expert from the Federal Environmental Agency and contributor to the Intergovernmental Panel on Climate Change, again reiterates the complexity of thinking about climate change which affects everything, saying 'even sneakers contained a greenhouse gas called SF6!' (2015). The majority white, European audience, representing countries like Bangladesh, Peru, the US and the Maldives, were then told if their emissions reduction commitments would be enough to keep global temperatures below the 2 degrees increase from pre-industrial levels. The juxtaposition of these positions, likely privileged Europeans making decisions representing a developing nation, replays some of the colonial (and postcolonial) power structures. According to Rimini Protokoll, more than 9000 people experienced the performance. They have also developed a school version with the Royal Meteorological Society that ties into the UK curriculum, called *Climate Change Negotiations for Schools* (with a resource pack available for free on the Royal Meteorological Society website).

World Climate Change Conference required an active engagement and participation from the audience: from moving between sites, to engaging in workshops, experiencing different scenarios and negotiating their delegation response to the climate change. For Baz Kershaw this variety of audiences responses can lead to a more relevant and democratic theatre. He argues that the discipline structures of a British or 'western' theatre have disempowered audiences, producing a political weakening due to 'an increasing severance from the wider environment (cultural, political, ecological) as a critical flexibility of responses was discouraged' (2007: 184). As Western theatre creates its own ecosystem, where theatre audiences are a niche, it separates itself from pressing political, cultural and ecological contexts and contributes to its own crisis of irrelevancy. He suggests that 'a reconnection of performance, community, politics and environment through theatre' (184) is needed, achieved by 'unruly audiences' who do not acquiesce to the hierarchical systems of cultural dominance produced by the disciplinary procedures. Kershaw suggests that 'creative promiscuity' in audience responses is needed to empower audiences: 'Theatrical performance is the most public of all the arts because it cannot be constituted without the direct participation of a public. Reduce that participation in any way and you reduce the environmental, the political and, especially, the democratic potential of theatre' (205). For Angelaki, *World Climate Change Conference* tapped in to this potential: 'the negotiations, deliberations and commitments that led to the delegations' declarations of intent, submitted by each group and summed up in the closing plenary, reinforced democracy as a tool not only for action but also for spectatorship' (2019: 44). A wide range of participatory responses and unruliness, then, can facilitate the environmental, political and democratic potential of theatre and performance practices, such as the engagement and negotiations that formed the ecodramatugy of *World Climate Change Conference*.

The structure of the performance facilitates audience members to take on different perspectives, as they are able to exercise agency within the performance which they cannot when watching often fruitless global negotiations. They are not asked to be a silent, applauding audience, rather they are asked to debate and negotiate. Through their roles, they are able to diversify their geographical perspective and get a sense of how a life in Germany may be connected to another life in Bangladesh. They can also flex the imaginative capacity of performance to test out utopian ideas if they want to. In staging this event simultaneously with COP20 in Lima, before the larger COP21 in Paris (and staging it again in 2015 to coincide with COP21), it acted as an alternative way to engage, where ordinary citizens find out more about the complex effects of climate change and could have a say in global climate change agreements. This kind immersive performance structure may 'produce participants in ecologically responsive action which recognises and embraces complexity in the agency of environments' (Kershaw 2007: 317). Participatory performance practices which embrace and lean into complexity in audience behaviour can be positioned as a way of better understanding the interrelationships between humans and nature, politics, capitalism and ecology, the Global North and the Global South (colonialism and migration), environment and performance.

Creative responses in participation and physical or bodily immersion in complex relations can be ecodramaturgical strategies for empowering audiences in ecological issues and revealing ideologies of oppression. The different works discussed in this chapter all acknowledge bodies as ecological and embedded in complex ecological relationships. Bodily ecologies, from the embodied experience of power relations in *World Climate Change Conference;* the myths and stories of wolves in *Wolf;* gendered and racialised bodies blending into landscapes in *Still Dance*, Mendieta's *Siluetas* and Pollard's photography; the exposed Chicano/a agricultural workers in *Heroes and Saints* to Luna's GM maize seeds on her body, can foster sensitivity or attunement to the ecological world, drawing out the neocolonial, capitalist, ableist, gendered and racialised power relationships that are inherent in ecological relationships. Ecodramaturgies can provide a spatial-temporal fissure, reminding us of our immersive relationship to the more-than-human world and its differentiated effects. We are always bodies in exchange with the ecological world, at once both sensitive and sensible. This exchange also leaves us exposed—to animal vulnerability, to the pesticides sprayed on our food and perhaps more profoundly to the power relations and systems that structure our ecological relationships. These relations are intersectional, requiring diverse and multiple responses, one of them being bodily practices that situate us in the knotty entanglements with the more-than-human.

Notes

1. Some of this writing in this section formed the basis for a short essay Woynarski, Lisa. 2017. 'Ecological Health in Violeta Luna's *NK603: Action for Performer & e-Maiz*' in *Theatre Applied: Performing Health and Wellbeing*, eds. Baxter, V. and Low, K.
2. GMOs are broadly used to produce crop species that are drought-resistant, pesticide-resistant or have a specific appearance or nutritional value.
3. Corn is also important to many Indigenous nations and cultures in so-called North America, with many, many strains unique to the Southwest, great plains, Midwest and Eastern Seaboard.

4. This awarded amount was subsequently knocked down on appeal to $78 million or a new trial for punitive damages (Stevens 2018).
5. This is not to exclude trans people who are also made vulnerable and exposed.
6. *Make it Work/Le Théâtre des négociations* (2015) was a similar piece that imagined a global climate change conference. The three-day event in Paris involved 200 students participating in the climate change conference ahead of COP21, with a Make It Work framework by Sciences Po. They had to make decisions and reorganise global environmental politics, beyond the existing structures. The large-scale experiment was a collaboration between Bruno Latour, SPEAP (Programme d'expérimentation en art politique de Sciences Po), Philippe Quesne, Nanterre-Amandiers and raumlaborberlin architects.

References

Abram, David. 1997. *The Spell of the Sensuous: Perception and Language in a More-Than-Human World*. 1st Vintage Books Ed. New York: Vintage Books.

———. 2011. *Becoming Animal: An Earthly Cosmology*. New York: Vintage Books.

Alaimo, Stacy. 2010. *Bodily Natures: Science, Environment, and the Material Self*. Bloomington: Indiana University Press.

———. 2013. Jellyfish Science, Jellyfish Aesthetics: Posthuman Reconfigurations of the Sensible. In *Thinking with Water*, ed. Cecilia Chen, Janine MacLeod, and Astrida Neimanis. Montreal, Quebec: McGill-Queen's University Press.

———. 2016. *Exposed: Environmental Politics and Pleasures in Posthuman Times*. Minneapolis: University of Minnesota Press.

Angelaki, Vicky. 2019. *Theatre & Environment*. London; New York: Red Globe Press.

Bennett, Jane. 2010. *Vibrant Matter: A Political Ecology of Things*. Durham: Duke University Press.

Best, Susan. 2007. The Serial Spaces of Ana Mendieta. *Art History* 30 (1): 57–82.

Chaudhuri, Una. 2012. The Silence of the Polar Bears: Performing (Climate) Change in the Theatre of Species. In *Readings in Performance and Ecology*, ed. Wendy Arons and Theresa J. May, 45–57. New York: Palgrave Macmillan.

———. 2014. Animal Acts for Changing Times, 2.0: A Field Guide to Interspecies Performance. In *Animal Acts: Performing Species Today*, ed. Una Chaudhuri and Holly Hughes, 1–12. Ann Arbor: University of Michigan Press.

———. 2016. Knowing Animals Now: The Unreliable Bestiary, a Multi-Part, Ongoing Performance Project by Deke Weaver. In *Reading Contemporary Performance: Theatricality Across Genres*, 183–184. London; New York: Routledge.

Clare, Eli. 1999. *Exile and Pride: Disability, Queerness, and Liberation*. Boston: South End Press.

Di Chiro, Giovanna. 2004. Producing "Roundup Ready" Communities? Human Genome Research and Environmental Justice Policy. In *New Perspectives on Environmental Justice: Gender, Sexuality, and Activism*, ed. Rachel Stein, 139–161. New Brunswick; London: Rutgers University Press.

———. 2010. Polluted Politics? Confronting Toxic Discourse, Sex Panic, and Eco-Normativity. In *Queer Ecologies: Sex, Nature, Politics, Desire*, ed. Catriona Mortimer-Sandilands and Bruce Erickson, 199–230. Bloomington: Indiana University Press.

Diamond, Cora. 2008. The Difficulty of Reality and the Difficulty of Philosophy. In *Philosophy and Animal Life*, ed. Stanley Cavell, Cora Diamond, John McDowell, Ian Hacking, and Cary Wolfe, 43–90. New York: Columbia University Press.

Drew, Jonathan. 2018. US Proposes Shrinking Last Endangered Red Wolf Habitat. *Phys.Org*, June 27. https://phys.org/news/2018-06-endangered-red-wolf-habitat.html.

Dunbar-Ortiz, Roxanne. 2014. *An Indigenous Peoples' History of the United States*, ReVisioning American History. Boston: Beacon Press.

Finney, Carolyn. 2014. *Black Faces, White Spaces: Reimagining the Relationship of African Americans to the Great Outdoors*. Chapel Hill: The University of North Carolina Press.

Foucart, Stephane, and Stephane Horel. 2017. « Monsanto Papers », Désinformation Organisée Autour Du Glyphosate. *Le Monde*, sec. Planet, October 4. https://www.lemonde.fr/planete/article/2017/10/04/monsanto-papers-desinformation-organisee-autour-du-glyphosate_5195771_3244.html.

Garner, Stanton B. 1994. *Bodied Spaces: Phenomenology and Performance in Contemporary Drama*. Ithaca: Cornell University Press.

Haraway, Donna. 2003. *The Companion Species Manifesto: Dogs, People, and Significant Otherness*. Paradigm 8. Chicago: Prickly Paradigm Press.

Ingold, Tim. 2000. *The Perception of the Environment: Essays on Livelihood, Dwelling and Skill*. Reissue. London; New York: Routledge.

Iovino, Serenella, and Serpil Oppermann, eds. 2014. *Material Ecocriticism*. Bloomington: Indiana University Press.

Johnson, Mark. 2007. *The Meaning of the Body: Aesthetics of Human Understanding*. Chicago; London: University of Chicago Press.

Kershaw, Baz. 2007. *Theatre Ecology: Environments and Performance Events*. Cambridge: Cambridge University Press.

LaDuke, Winona. 1999. *All Our Relations: Native Struggles for Land and Life*. Boston: South End Press.

Langston, Nancy. 2003. Gender Transformed: Endocrine Disruptors in the Environment. In *Seeing Nature through Gender*, ed. Virginia J. Scharff, 129–168. Lawrence: University Press of Kansas.

Luna, Violeta. 2014a. *NK 603: Action for Performer & e-Maiz*. Performance. Accessed April 03, 2019. http://www.violetaluna.com/NK603.html.

———. 2014b. *Body in Action: Cartographies for Socially Engaged Performance*. Keynote, Wellington, New Zealand: Australasian Association for Theatre Drama and Performance Studies (ADSA), June 27.

May, Theresa J. 2007. Beyond Bambi: Toward a Dangerous Ecocriticism in Theatre Studies. *Theatre Topics* 17 (2): 95–110.

Mendieta, Ana, and Bonnie Clearwater. 1993. *Ana Mendieta: A Book of Works*. Miami Beach: Grassfield Press.

Monsanto. 2002. Safety Assessment of Roundup Ready Corn Event NK603. Accessed January 28, 2015. http://www.monsanto.com/products/documents/safety-summaries/corn_pss_nk603.pdf.

Moraga, Cherríe. 1993. *The Last Generation: Prose and Poetry*. Boston: South End Press.

———. 1994. *Heroes and Saints and Other Plays*. Albuquerque: West End Press.

Mortimer-Sandilands, Catriona, and Bruce Erickson, eds. 2010. *Queer Ecologies: Sex, Nature, Politics, Desire*. Bloomington: Indiana University Press.

Muñoz, José Esteban. 2011. Vitalism's after-Burn: The Sense of Ana Mendieta. *Women & Performance: A Journal of Feminist Theory* 21 (2): 191–198.

Pascoe, Bruce. 2014. *Dark Emu: Black Seeds: Agriculture or Accident?* Broome, Western Australia: Magabala Books.

Pollard, Ingrid. 1987. *Pastoral Interlude*. Photograph. Accessed March 23, 2020. https://collections.vam.ac.uk/item/O107865/pastoral-interludeits-as-if-the-photograph-pollard-ingrid/.

Poynor, Helen. 2009. Anna Halprin and the Sea Ranch Collective, an Embodied Engagement with Place. *Journal of Dance & Somatic Practices* 1 (1): 121–132.

Rimini Protokoll. 2014. *WELT-KLIMAKONFERENZ (World Climate Change Conference)*. Performance. Deutsches Schauspielhaus Theatre, Hamburg

———. 2015. *WELT-KLIMAKONFERENZ (World Climate Change Conference) | Helgard Haug, Stefan Kaegi, Daniel Wetzel (with English Subtitles)*. Accessed April 08, 2019. https://vimeo.com/129199465.

Rothfels, Nigel. 2014. Commentary: A Hero's Death. In *Critical Performances: Animal Acts: Performing Species Today*, ed. Una Chaudhuri and Holly Hughes, 182–188. Ann Arbor: University of Michigan Press.

Shiva, Vandana. 1997. *Biopiracy: The Plunder of Nature and Knowledge*. Boston: South End Press.

Stein, Rachel. 2004. *New Perspectives on Environmental Justice: Gender, Sexuality, and Activism*. New Brunswick; London: Rutgers University Press.

Stevens, Richard G. 'Bugs'. 2018. Roundup Weed Killer Lawsuit Hits a Snag, but Monsanto Is Not off the Hook. *The Conversation*, October 26. http://theconversation.com/roundup-weed-killer-lawsuit-hits-a-snag-but-monsanto-is-not-off-the-hook-105559.

Stubblefield, Eeo. 1997. *Still Dance with Anna Halprin.* Dance. Accessed May 03, 2019. http://www.stilldance.net/.

———. n.d. Eeo's Statements. Accessed April 5, 2019. http://www.stilldance.net/Statements01.htm.

Thomas, Arden. 2012. Stillness in Nature: Eeo Stubblefield's Still Dance with Anna Halprin. In *Readings in Performance and Ecology*, ed. Wendy Arons and Theresa J. May, 113–126. New York: Palgrave Macmillan.

Weaver, Deke. 2013. *Wolf.* Performance, Champaign, Illinois, September 2013.

———. n.d. The Unreliable Bestiary. Accessed April 8, 2019 https://www.unreliablebestiary.org/.

Weiss, Gail. 1998. *Body Images: Embodiment as Intercorporeality.* New York: Routledge.

Whitehead, Simon. 2018. Louphole: Drifting with Wolves. *Performance Research* 23 (7): 120–121.

Wolfe, Cary. 2014. Commentary: Apes Like Us. In *Critical Performances: Animal Acts: Performing Species Today*, ed. Una Chaudhuri and Holly Hughes, 156–162. Ann Arbor: University of Michigan Press.

Woynarski, Lisa. 2017. Ecological Health in Violeta Luna's NK 603: Action for Performer & e-Maiz. In *Theatre Applied: Performing Health and Wellbeing*, ed. Katharine Low and Veronica Baxter. London: Bloomsbury Methuen Drama.

CHAPTER 5

Earth As Home: Local, Imagined, Material, Global

A rainy date in September in Cardiff, Wales has people huddled under plastic sheets and rain jackets to watch the performance happening inside the *Trans-Plantable Living Room*. Under a large black walnut tree in Bute Park an extensive set of vintage and recycled living room furniture has been installed. The set has been planted with colourful, edible plants, flowers and herbs, all poking up unexpectedly from ordinary household items. Picture frames hang from one of the branches, framing the surrounding park as artworks in the space. A yellow wall clock hangs on the trunk of the tree. A boxy, grey hollowed-out television grows sprouts. Tables made of large wood spools have been planted with edible flowers and woven with willow. Wooden chairs have been fixed to a base where orange and red nasturtiums are blossoming, while the seats of other chairs have been planted up with lush greens. A 'chandelier' of yellow and red chilli peppers hangs over the set. Herbs and flowers are growing from dresser drawers, the corners of armchairs and sofas, door frames, coat racks, tea cups, handbags, books and rain boots. Passers-by, dog walkers, children, and people of all ages walking through the park to get from one side of the city to another, stop at the novelty of the installation. They take photos and are invited to sit down and have a cup of tea made with some of herbs grown in the living room.

The Trans-Plantable Living Room was a living, edible performance space, grown by community gardeners, that hosted interactive performances by performance collective Plantable (of which I am a member) in

L. Woynarski, *Ecodramaturgies*, New Dramaturgies,
https://doi.org/10.1007/978-3-030-55853-6_5

September 2013 in Cardiff in Bute Park behind the Royal Welsh College of Music and Drama, as part of the World Stage Design festival.

The domestic set of the *Trans-Plantable Living Room* was an evocative image of home. The idea of home is central to ecological discourse as the Greek root of 'eco' is *oikos*, meaning home or dwelling place. According to Seamon and Mugerauer, dwelling is what turns a place into home and 'involves the process by which a place in which we exist becomes a personal world or home. Dwelling incorporates environments and places but extends beyond them, signifying our inescapable immersion in the present world as well as the possibility of reaching beyond to new places, experiences and ideas' (1985: 8). Dwelling is based on the idea of immersion in the more-than-human world and considers the way in which we live immersed and imbricated in ecological relationships. This chapter theorises the way in which ecodramaturgies can extend and nuance conceptions of home and place that include a globalised perspective. In the context of the current ecological crisis, 'we must ask what human dwelling on earth is and how it is possible to have a home' (Seamon and Mugerauer 1). In other words, dwelling asks how we live on earth, where we live and how we think about it in relation to home. Climate change knows no borders and calls into question the localism of home as Blunt and Dowling note: 'home on a domestic scale is inextricably bound to national and imperial geographies of home' (Blunt and Dowling 2006: 51). Extending these questions of how and where we live contributes to an understanding of our place within the mesh of ecological relationships.

In this chapter, I argue that place-based practices integrate ideas of home, drawing out the ways in which is it understood as ecological, material, imagined and global, inflected through race, gender, colonialism and migration. I begin my discussion with *Trans-Plantable Living Room* and Fevered Sleep's *Above Me the Wide Blue Sky* (2013), analysing the politics of home and place as ecological. Then I turn to the American classic *A Raisin in the Sun* (1959) by Lorraine Hansberry and the 'homeplace making' (hooks 1990) in the play which I read as ecologically material as well as shaped by environmental racism. The determination of the characters to create a home where they can dig in the dirt and affirm their humanity mirrors environmental justice work that positions home making as an act of resistance to the racist socio-political forces that shape their ecological relationships to home. I then extend the concept of home to global place perspectives, or eco-cosmopolitanism (Heise 2008) in the installation

#crazyweather (2013) and the play *Carla and Lewis* (2014) by Shonni Enelow, where the tension between the local and global is interrogated.

Oikos as Home in *Trans-Plantable Living Room* and *Above Me the Wide Blue Sky*

Home is a complex and contested idea. For Una Chaudhuri, theatre can be considered a type of home: 'a specific place to be inhabited for a specific time and in a special way: not exclusively, not through ownership, but for the direct sharing of experience' (1995: 83). For dance artist Simon Whitehead the body is type of home:

> the body is the 'first home', and the place or the territory where we live is a 'second home', or perhaps something that is made through our heightened sense of awareness to where we are. The 'third' home is the home you discover when you start interacting with the assemblage of body and environment to produce something new (in Lavery and Whitehead 2012: 114).

There are many nuanced ways of thinking about the relationship between eco/*oikos* as home. Ecology as home may be closer to what Harré et al. refer to as the 'Australian Western Desert concept of *nguaia* (camp, home, country, place where people are staying or could stay)' (Harré et al. 1998: 106). Blunt and Dowling consider home on different scales, arguing that the idea of home is often not confined to a house and may be extended to 'suburb, neighbourhood, nation, or indeed the world' (2006: 29). Extending the idea of home, based on understanding our ecological enmeshments, including socio-political forces and structures, is the focus here.

Different aspects of home and place are manifested, amplified and critiqued in my practice-based experiment the *Trans-Plantable Living Room* and Fevered Sleep's *Above Me the Wide Blue Sky*. Domesticity, nostalgia and geopathology all shape the idea of home in an ecological sense, at times reductively and at others generatively. Domesticity and nostalgia are concepts inherent in ideas of home. Home as a domestic space, or the distinct boundaries between the public and private, is also an ethnocentric concept, according to Saunders and Williams (1988), as it has become a Western cultural symbol. 'Nostalgia' as Mackey and Whybrow point out, 'literally splits…into the pain suffered (*algos*) as a consequence of being unable to fulfil the dream of return (*nostos*) to a location perceived to be "home"' (2007: 2–3). Nostalgia is the pain of returning, or longing for, a home that cannot be reached. It has implications of temporality and the past, as well as being implicitly regressive and backwards looking, in some contexts.

Trans-Plantable Living Room,[1] on which I was a collaborator, was conceived to celebrate urban community food growing and to engage local groups of growers in a creative performance project. It involved three performance-makers (from the UK, Canada and US), an Australian ecoscenographer, a community allotment with 150 volunteers (in Cardiff) and 17 oral history interviews with community gardeners conducted by a community-engagement coordinator Rosie Leach, to gather personal narratives about gardening in Cardiff, collecting stories of people involved in Riverside Community Allotment. This included asking why people garden, how gardening practices have changed over time in relation to climate change and what role they see local, small-scale food production taking in providing food in today's climate-changing world. Based on a Living Stage design first realised in Australia by designer and ecoscenographer Tanja Beer, the project was part growing experiment, part tea party and part performance. The outdoor living room filled with plants was grown by Riverside Community Allotment and a network of local gardeners. Ideas and material from workshops and interviews were woven together by Plantable Performance Research Collective (Bronwyn Preece, Meghan Moe Beitiks and myself), to create a performative interaction with the space. Excerpts from the interviews were edited into a soundscape that shaped the performance (Fig. 5.1).

Fevered Sleep's *Above Me the Wide Blue Sky* was both a performance and installation at the Young Vic Theatre, London. A single performer recalled collections of stories and descriptions gathered from people about their ideas of home and the changing landscape of that home. Laura Cubitt, the white female-presenting performer, described different experiences, images, tastes and smells of the more-than-human world in a stream of consciousness style meditation (accompanied by a dog). These experiences and images included passages such as 'There are boys kicking a curled-up hedgehog like a football', 'The taste of coal', 'The deep black abyss of an infinite, star-studded night' and 'The smell of rain falling on hot tarmac' (2013). Videos of slowly shifting skyscapes were presented on all four walls of the theatre, encompassing the audience who sat in the round in the black box space, while the performer stood on chalk tiles in the middle of the room or sat on a revolving wooden stool. Cylindrical lights on poles of different heights were spread out on stage producing

Fig. 5.1 Plantable (Lisa Woynarski, Megan Moe Beitiks, Bronwyn Preece), *Trans-Plantable Living*, Cardiff, Wales, 2013. (Photo: Nigel Pugh, courtesy of the artists)

softly glowing light that pulsed at different times underscoring the text, while the dog slept soundly on a blanket beside the performer (Fig. 5.2). A soundtrack of birdsong and string instruments and electronic effects accompanied the text and imagery, building and changing. Film projectors lit up some of the upturned chalk bricks with small images of flowing water. The installation was open before the performance start time for audience members to experience (without the performer or dog). The script was a series of fragmented memories, stories and descriptions of interactions with the environment, slow moving and contemplative. The company website describes it as:

> A collection of stories about our deep-rooted, deeply felt, easily overlooked and profoundly important connection to the land, the sky, the sea, the weather, and the other living things that surround us (Fevered Sleep 2013).

Halfway through, the script reversed as dark clouds covered the walls. The performer recounted the memories in reverse order, changing the present

Fig. 5.2 Fevered Sleep, *Above Me the Wide Blue Sky*, Young Vic, London, 2013. (Photo: Matthew Andrews, courtesy of Fevered Sleep)

tense to the past tense 'There used to be…', suggesting loss and nostalgia for the natural world being portrayed.

Both performances contained different ideas of home in relation to a privileged idea of home and mobility. Home was extended beyond a bounded material locality and understood as being made up of what Blunt and Dowling call 'complex [eco-]socio-spatial relations and emotions' (Blunt and Dowling 2006: 3). Blunt and Dowling identify three components of a critical geography of home: 'home as simultaneously material and imaginative; the nexus between home, power and identity; and home as multi-scalar' (Blunt and Dowling 22). All three of these aspects have resonances with the idea of earth as home, such as a global sense of place as intersection point (Massey 1997), climate change refugees and environmental racism (Huggan and Tiffin 2010), and home on different scales (Tuan 1974). Massey (1997) suggests that a global sense of place is influenced by race, gender and social relations, as well as capitalism, which affects the so-called global mobility that characterises contemporary living, underscoring how home is shaped by power and identity. The *Trans-Plantable Living Room* was transnational in scope, with people travelling from Australia, US and Canada to take part in making it. This is a very privileged sense of global mobility. The ability to live, work or study on an international scale is the domain of a social class with the means to do so; it is not necessarily out of necessity or due to lack of a home. My sense of global accessibility (and a transnational home) is very different from a refugee, asylum seeker or economic migrant. These are very different forms of deterritorialisation. I am mindful that we therefore approached the idea of earth as home from a privileged position.

Above Me the Wide Blue Sky and *Trans-Plantable Living Room* addressed the perceived disconnection between the contemporary Western context and understanding ourselves as part of (always already immersed in) the more-than-human world. This is a form of Chaudhuri's concept of geopathology or problem of place: 'a series of ruptures and displacements in various orders of location, from the micro-to the macrospatial, from home to nature, with intermediary space concepts such as neighbourhood, hometown, community, and country ranged in between' (Chaudhuri 1995: 55). For Chaudhuri, the principal dislocation to which geopathology refers is the dislocation between humans and the natural world, which she contends is manifested in theatre through the reduction of 'nature' to scenery or setting (1995: 55) rather than recognising the agentic capacity of the more-than-human. Accordingly, home then is both a material container for identity and a concept of entrapment, which leads to the 'desire to deterritorialize the self' (Chaudhuri 82). Identity is constructed not in relation to a material place but rather in relation to the enmeshment of living relationships, which are shifting and dynamic, not static or rooted.

Both performances focused on the more-than-human (plants, a dog, sky, landscapes) in an effort to draw attention to place-based relations between humans and more-than-humans. Chaudhuri characterises geopathic disorders as 'the suffering caused by one's location' (1995: 58) and contends this contributes to the anthropocentrism of theatre. For Chaudhuri, the 'first decisive act' (the clearing of the forests in order to build Rome, the first city and empire), was the first version of home in the Western cultural imagination, causing the perceived separation between humans and the natural world (the former having dominion over the latter). This foundational idea of home 'is predicated on the destruction of the woods. It brings with it both stability and loss' (Chaudhuri 75). Home is made up of a duality between a rooted stability in a material location or a stable identity and a kind of oppression or loss. Stability and loss are interesting to consider in light of ecological vulnerability and global climate change.

Understanding the potential for ecological loss affects our sense of home and our place within ecological relationships.

About halfway through *Above Me the Wide Blue Sky,* the performer told a story about how the rural landscape of her childhood home had changed over the years with the loss of skylarks. 'When I think of home, I think of a field' (Fevered Sleep 2013). She goes on to describe the field in detail, noting how the landscape has changed with more cars on the road, less birds. She then repeats the descriptions from the first half of the show in reverse order, changing them to the past tense, while the sky turns dark and black clouds roll in. 'There used to be birdsong...There used to be flowers... There used to be skylarks' (2013). The darkness invoked by this imagery referenced a degrading planet, the effects of climate change and the resulting disasters. Fevered Sleep were interested in the question: 'What do we really lose if we lose our connections to the non-human world that surrounds us?' (2013). This is a question of geopathology or dislocation from nature; however, it also implies that nature is a backdrop to human action, which may surround us, but we can 'disconnect' from it. Rather than conceiving of the more-than-human world as something that surrounds us externally, to resist the tacit anthropocentrism, I understand it is a mesh of relationships in which we are always embedded, whether we sense and see the interconnections or not. The answer to geopathology is not creating 'connections to the non-human world', but rather it is about understanding and making visible those already existing relationships.

In the *Trans-Plantable Living Room,* we sought to resist or problematise the anthropocentrism and romantic idealism inherent in performance by acknowledging the agency of the plants as co-performers in the piece, enacting a bioperformativity. As described in Chap. 3, following ecomaterialism, the world is alive with actants, operators and human–non-human assemblages. *Above Me*, by calling attention to the sensuous experience of being immersed in the more-than-human world, creates what Bennett calls 'a cultivated, patient, sensory attentiveness to nonhuman forces operating outside and inside the human body' (Bennett 2010: xiv). In this way, the work made the patterns and relationships of the ecological world visible, drawing out the ways in which we are immersed in a multitude of sensuous exchanges (from the shifting clouds, to birdsong, to badgers, to rain on tarmac). *Trans-Plantable Living Room* was based on a tea party pastiche, but it was a tea party for plants. We subverted the domestic imagery by chewing on planted parsley in tea cups instead of drinking tea, hosting the plants instead of the audience and then, at the end of performance, we gave out tea bags filled with spinach seeds for the audience to take home and plant. The soundtrack we used in the performance was of the community gardeners' voices, speaking about their experience of gardening today which included adjusting to the effects of climate change. The interview clips used were not backwards-looking and did not valorise the past as a time of ecological balance. On the conceptual level, the living room communicated the idea that we are embedded and immersed in the more-than-human world: by setting it outdoors, around a tree, and modifying the furniture to house the plants, it communicated the aliveness of our living room and decentred the human by focusing on the plants. Although acknowledging the romantic and anthropocentric implications of the piece, we hoped to sufficiently disrupt them to reframe the idea of home within an ecological context.

Both *Above Me the Wide Blue Sky* and *Trans-Plantable Living Room* evoked a nostalgia in a comfy, affectionate and mostly untroubled version of home. Interpreting the idea of oikos as home, philosopher Roger Scruton (2012) describes 'oikophilia' as the love of home. According to Scruton, it is oikophilia, or eliciting a love of home, that will bring solutions to the ecological crisis. It is his contention that in order to care for or conserve something you must first love it, and that through local, small-scale initiatives motivated by a love of one's own place the ecological crisis will be effectively addressed. However, oikophilia can be dangerous in its potential to be exclusionary, invoking what Mackey and Whybrow refer to as the negative practices of local place, including a 'retrospective idealism of nostalgia' that leads to 'territorialisation, introversion, defensiveness and boundary-making that excludes difference' (2007: 6). Scruton's focus on nostalgia and the past romanticises the rural English village (which is the context in which he is writing) with nationalistic and xenophobic undertones, often coded as white, by creating place-based identities that are necessarily exclusionary to 'outsiders'.

In order to be effective, oikophilia would need to be extended beyond just love of a local place. Otherwise, an individualist logic would allow for the abdication of responsibility for far-reaching or distance ecological effects, as long as one's own local area was 'clean', safe from pollution and other harmful environmental effects. Following this logic, we would only need to care or take responsibility for our immediate area (as Scruton's conception of home is a local village) without understanding or acknowledging the way in which our lives are entangled in a number of ecological relationships spanning the global, from our clothes, food, rubbish, among other things. This solipsistic view is already present in much waste management, with rubbish being shipped to other countries as if once it crosses a geographical border it is no longer a problem, or that wealthy nations can pay to shift their waste onto those without the wealth or political capital to effectively resist it. This is also evident in the lack of concern/action over the large floating islands of plastic waste in the oceans, such as the Great Pacific Garbage Patch. Ostensibly, these are not a problem because they are not considered anyone's 'home' (or not of human value). Scruton argues that localised, decentralised organisations are much more effective than larger systems of governing (he identifies as a Conservative environmentalist). If responsibility was local, he suggests, we could make a lot more ecological progress in environmental management and climate change mitigation than within our current system in the UK. However, this can reinforce dangerous nationalism and 'hard borders', with poor communities and communities of colour facing the brunt of the pollution from waste management or power generation. This emphasis on localism can cause an exclusionary 'us versus them' mentality (Harvey 1996; Massey 1997). In addition to environmental racism, Harvey uses the proliferation of gated communities in America to illustrate one of the potential outcomes of this kind of localism.

The shift in text in *Above Me the Wide Blue Sky* from describing things as 'There is' to 'There used to be' contributed to a sense of ecological nostalgia which can be misrepresentative of history. As employed by Scruton (2012) and many others, ecological nostalgia seems to be referencing an unspecified past, where people were more connected to the land and lived in balance with the natural world. When invoked, this seems at best wishful thinking, and at worst regressive rhetoric. Of course, there are many ecological lessons to be learned from specific historical moments and many industries have sought them out (such as the popularity of organic farming, cooking 'from scratch' and growing your own vegetables). However, a blanket nostalgia or assumption of the inherent merits of the past ignores the contested and oppressive histories of home, domesticity and labour, as well as the ecological problems of the past. In *Above Me the Wide Blue Sky*, the shift from present to past tense gestured to loss of the 'better' past:

> When I think of home, I think of the field, the field as it used to be. I think of the conkers, and the rooks and the direction of the wind that carries the

sound of the motorway, which is louder than it used to be. Everything is different than how it used to be. (Fevered Sleep 2013)

The backwards-looking focus of the second half of the performance emphasises loss of 'nature'. There are several references to fields and green hills such as 'snow falling on a ploughed field' (Fevered Sleep 2013), employing the trope of Britain as a 'green and pleasant land'. Ecological nostalgia, evoked by a relationship to nature as scenery through references to fields, hills, children playing and trees, could be read as re-inscribing the myth of the rural landscape as a 'pure space' or rural idyll, which cannot accommodate difference. Cloke et al. (2000) contend there is a close relation between the idea of 'home' and rurality, prevalent not only in Britain but also in rural America (among other places) with its iconography of 'mom and apple pie' (Cresswell 2004: 114). They argue that the image of home as a rural idyll valorises a particular and normative idea of the home and domesticity, one that is ostensibly white and middle class. Belonging to specific local places can reinforce a reductive and problematic idea of 'traditional community', which invokes troubling nostalgia leading to oppressive and segregated areas, unwelcoming to 'outsiders'. This kind of localism can produce areas that are 'oppressive', which ghettoise those areas outside of the gates (Harvey 1996) or walls. This idea of home also has alluring and comforting qualities that can make it aspirational. However, the longing or nostalgia for this particular image of home can necessarily 'other' anything non-normative.

Ideas of domesticity and nostalgia were also present in *Trans-Plantable Living Room*. The domesticity of the living room, which we referred to as 'grandmother's living room', had a quality of nostalgia about it. The floral print wing chair, the crocheted doilies, the china tea cups and mismatched saucers, the straw hats hanging on the hat rack, the old television hollowed out and planted with sprouts, framed pressed flowers, greeting cards with pictures of older women, all contributed to both the domestic and nostalgic aspects of the space. This echoes Irene Cieraad's (1999) conception of domesticity being intertwined with nostalgia in understandings of home: 'the concept of domestic space introduces not only an inevitable historical dimension, but also a temporal dimension often clad in nostalgic images. The idea of domesticity is one of the most powerful images related to domestic space' (3). These images of domestic nostalgia could be read as reinforcing a romantic and regressive idea of 'nature', backward-looking and suggestive of a 'return to nature' position, which although not our intention in making the piece, was referenced in the scenography and tea party format.

The idea of home as an ideal and longed-for place of security or 'authentic' place, as suggested by the concept of oikophilia, is not a universal conception of home, as feminist thinkers like Gillian Rose (1993) remind us. Home and its spatial makeup historically elicited sexism and oppression. For many people, home is not a secure haven free from conflict, it can be the primary place of oppression, therefore problematising the 'claim that home provides the ultimate sense of place' (Rose 47). For many, home can be a source of distress, conflict, trauma, displacement or a sense of homelessness. Home is not an uncontested nurturing place of security, complicating oikophilia as a gendered, raced and class-based conception.

The domesticity and nostalgia of both *Above Me the Wide Blue Sky* and *Trans-Plantable Living Room* also gendered the space. This was reinforced by the fact that the core artistic team for *Trans-Plantable Living Room* were all white and female-presenting, including the three performers, and the performer in *Above Me the Wide Blue Sky* was also white and female-presenting (with a writing team of two white people). The concept of 'home' in both pieces was, in a sense, dangerously nostalgic when thinking about climate change. Such domesticity could be read as a regressive sentiment, a return to simpler times if we are to mitigate the effects of climate change. In the *Trans-Plantable Living Room*, we actively tried to disrupt this reading through the retrofitting of the furniture and through the interaction with the space in performance. The principles of the design in the piece were based on Beer's ecoscenography (an ecologically orientated scenography), which sought to understand the life of the set pieces beyond the performance: 'integrating an awareness that no decision stands on its own: every choice is intertwined with social, environmental, economic and political consequences that are far reaching and capable of having long term effects' (Beer n.d.).[2] The plants and furniture were returned to the community allotment and were used to set up a social space. We attempted to disrupt the 'cosy' aesthetic by slashing the upholstery of the chairs and sofa to house plants, removing the seats of the dinner chairs to plant them, cutting books open to house succulents; all of which was to disturb a domestic and nostalgic reading with the infusion of 'wild', outdoor elements. However, the plants were all 'cultivated' or perhaps even domesticated for the process of building the installation. This paradoxical quality characterises the tensions of the human relationship with the more-than-human.

With the *Trans-Plantable Living Room*, we considered gardening not only as a pragmatic mechanism of subsistence or domestic activity, but also as a site of ecological activism and resistance. Gardens are not without their own set of ethical imperatives, including who has access to these sites, what kind of plants are grown, power relations, gender roles and what kind of resources are consumed in their production. As Mackey has asserted, 'historically, allotments are places that are deeply contested' (2007: 184). One of the interviewees for the project identified the rising cost of rent for allotments as local councils make cuts to them: 'allotment rents have doubled in the last year, roughly, and they're going to double again next year' (Watts 2013). Community gardening as resistance, and home writ large, is further constrained and constructed by socio-political power formations.

Home reaches across different scales because, for Cresswell, 'by transforming the Earth into home we create places on a myriad of levels' (Cresswell 2004: 24). The concept that home can occur across large geographical expanses, spans of time or more intimate settings reflects that 'home' is imaginatively constructed, as a sense of belonging can be tied to different scales of place, from a park bench to a globalised, transnational space (Blunt and Dowling 2006: 29). Localism itself does not have clearly defined borders and can span different scales of time and space. Not just material places; images of home can be constructed through ideas of normativity, intimacy, the domestic, exclusion, oppression and private space. Thinking about the earth as home draws out some of the contested territory of home, much of which has ecological relevance. This shares conceptual territory with dark ecology,[3] as the earth (or 'nature') can be considered a site of conflict, trauma and distress. Across scales of home, from nation to park bench, the ecodramaturgies of performances such as *Above Me the Wide Blue Sky* and *Trans-Plantable Living Room* can reveal and refract these various scales and localities of home, prompting us to think differently about our relationship to them. Rather than oikophilia or localism, Massey's call for 'a need to build a "local" politics that thinks beyond the local…*against* localism but *for* a politics of place' (2007: 15) may be a more generative way of considering the enactment of home that accounts for the eco-socio-political structures involved.

Performance can highlight, reveal and critique ecological relationships. In so doing, it can also dissect and amplify ecological values and assumptions, considered here in relation to ideas of home. *Above Me the Wide Blue Sky* and *Trans-Plantable Living Room* manifest and expose a type of ecological nostalgia and geopathology, while also attempting to grapple with current ecological conditions. Critiquing these ideas extends my configuration of ecodramaturgies to understand the way in which theatre and performance elucidate and perform concepts of place, ecology and home.

Home and Environmental Justice in *A Raisin in the Sun*

The focus of readings of the canonical American play, *A Raisin in the Sun* (1959) by Lorraine Hansberry, has often been about the myth of the American Dream and how it is challenged by race, even at the time of supposed liberation. *A Raisin in the Sun* was the first play produced on Broadway by a Black woman and won numerous awards including the New York Critics' Circle Award. Much has been written about the play and its groundbreaking status in terms of representation, social realism and the racialised American Dream (Matthews 2008) but there has been less focus on it as an ecological play (with the exception of May 2006). *Raisin* contains multiple ideas of home across different scales (Matthews 2008), which I recontextualise in terms of ecodramaturgies, place and ideas of home: violent, imagined and ecological, drawing on intersectional ecologies that acknowledge the entanglement of race and class in concepts of home.

The racial violence of home, such as the 1946 bombing and riots by white people to keep Black people out of public housing in Chicago (Meyer 2001), is one of the portrayals of home in the play. When matriarch Lena Younger decides to buy a home in a predominantly white neighbourhood, Clybourne Park, with the insurance money from the death of her husband, the family face resistance from the white residents. Upon hearing the news of the move, Lena's daughter-in-law, Ruth, is astonished because 'there ain't no coloured people living in Clybourne Park' (Hansberry [1959] 2009: 70). Mama's response is that she tried to buy the nicest place for the least amount of money and 'them houses they put up for coloured in them areas way out all seem to cost twice as much as other houses' (71). Just as the Youngers are intimidated to leave Clybourne Park by Mr. Lindner, a representative from the new neighbourhood association, 'resistance to neighbourhood integration happened on a smaller scale, with blacks being denied home loans, paying higher finance cost (if, indeed, they received a loan), and facing federally supported racially restrictive zoning ordinances and covenants' (Matthews 2008: 556). The Youngers' neighbour, Mrs. Johnson, comments on what kind of reception the Youngers might receive in their new neighbourhood. She mentions a recent newspaper story about 'coloured people that was bombed out their place' and later says 'I bet this time next month y'all's names will have been in the papers plenty—"NEGROES INVADE CLYBOURNE PARK—BOMBED!"' ([1959] 2009: 76–77). The storyline was based on the personal experience of Lorraine Hansberry as her family attempted to move to a white area in Chicago with a racial covenant (a restriction that non-white ethnic groups were banned from buying or leasing property there) in the 1930s and faced 'vigilante violence', they weren't forced out but sued (Matthews 2008). The case went to the Supreme Court which sided with Hansberry's father. The win was on a technicality though and failed to establish a legal precedent. Restrictive racial covenants were upheld until 1948 in America.

These restrictive housing codes are one type of environmental racism that limits access to home. Kristin Matthews points out that 'African Americans were denied subjectivity and humanity and therefore were excluded from the promises of "home": safety, shelter, ownership and citizenship' (Matthews 2008: 569). Benjamin Chavis outlines the different forms of environmental racism enacted and enforced including 'racial discrimination in environmental policymaking'; 'in the enforcement of regulations and laws'; 'in the deliberate targeting of communities of color for toxic waste disposal and the siting of polluting industries' and 'in the official sanctioning of the life threatening presence of poisons and pollutants in communities of color' (1992: 3). Environmental pollution is shaped not by poverty or income level but by race. One of the first studies in 1987 by the United Church of Christ, *Toxic Waste and Race in the United States*, found that the way to predict the presence of hazardous waste was through racial categorisation of communities, a better predictor than income levels or value of real estate (Westra and Wenz 1995: xv–xvi). This finding caused Chavis to develop the term 'environmental racism' to refer to the way in which social constructions of race and discrimination were dominant factors in shaping relationships to environment.

The Youngers live in Southside, Chicago, and Theresa J. May points out that 'the arc of the play follows a quest for a place that can support and nurture the family's material/ecological and emotional/spiritual well-being. May argues that 'Beneatha's diagnosis that her family suffers from "acute ghetto-itis" is more than a figure of speech; here, Hansberry speaks to the material/ecological situatedness of her characters' lives.' (May 2006: 130). The perception of urban areas and cities as being the home of people of colour has contributed to the dichotomy of cities as 'polluted' and suburbs as 'safe and clean'. Lawson (1995) claims that lack of attention to the city from mainstream environmentalism has to do with the racialised conception of place: urban living in cities with large populations of people of colour are considered 'unsafe' and 'polluted' while the predominantly white suburbs are often considered 'safe', 'clean' and 'pure'. If cities are already considered 'polluted' it becomes justifiable to have more dumping there, especially if the political capital of the residents is low and there will be little pushback. Lawson writes that 'because of the large black population these cities [Philadelphia, New York, Baltimore, Chicago, Atlanta] are often viewed as unsafe...There is a tendency to view some cities as black enclaves. Suburbs are more likely than not viewed as white enclaves. In this way, urban and suburban space has become racialized' (48). The urban migration of Black Americans, the Great Migration between 1916 and 1970, was to build safe homes away from the dangers of lynching and segregation. May points out the characters in *Raisin* were part of this migration: 'Hansberry's Lena and Walter, Sr., were part of the Great Migration ...[when] black families moved in great numbers to the urban North, seeking economic opportunities, freedom from segregation laws and the terror of lynching, and leaving behind the uncertainty of the sharecrop system' (May 2006: 131). This migration did not achieve the safe haven the families were seeking, at least not by 1959; rather, the systemic racism enforced a low standard of living in urban centres that were seen as enclaves of people of colour.

Laura Pulido argues that we must consider racism as socio-spatial in order to understand how it shapes places: it is 'both constitutive of the city and produced by it' (2000: 13). She argues that white privilege has been an influential factor in understanding and constituting relationships between places, as suburbanisation 'has largely been an exclusionary undertaking' (33). Writing about Black geographies, Katherine McKittrick contends that 'in the Americas, it is impossible to delink the built environment, the urban and blackness' (2013: 2). Using a recovered burial site for enslaved people in New York City, she argues that sites of the city demonstrate that 'the geographies of slavery, postslavery and black dispossession'

(McKittrick 2) mean that the act of survival is spatialised. McKittrick uses plantation logic to uncover the ways in which the contemporary city has shaped black violence, questioning the humanity of the Black people who live there: 'plantation—precisely *because* it housed and historicizes racial violences that demanded innovative resistances—stands as a meaningful conceptual palimpsest to contemporary cityscapes that continue to harbor the lives of the most marginalized' (McKittrick 5). Plantation logic informs the city because it also uses 'black forced labor' to mobilise 'an economic structure that would underpin town and industry development in the Americas. With this in mind, the plantation spatialises early conceptions of urban life within the context of a racial economy' (McKittrick 8). This can be extended to migrant workers, who are compelled to build homes they can never afford to live in, or who work on farms, exposing themselves to harmful pesticides, having to trade their health for their livelihood.

Lena Younger seeks to leave a home in an urban centre where she sees her family being dehumanised by the plantation logic of the city and she is exposed 'to hazardous materials and toxic wastes, pollution, health hazards, and resource depletion' (Lawson 1995: 49). Instead, she strives to find a home where her family has 'safe living space, a safe workplace, clean water and air, and the political power to influence environmental decisions' (Lawson 49) which she believes will be found in the white neighbourhood of Clybourne Park. The Southside apartment in *Raisin* is described as having only one window, 'the sole natural light the family may enjoy in the course of a day is only that which fights its way through this little window' ([1959] 2009: 9). May describes the Youngers' apartment as 'a habitat on the edge. Despite her best efforts, Lena's home is increasingly threatened and unsustainable: space is inadequate; furnishings, tempers, and relationships fray; a woman contemplates abortion because there is no room for another mouth to feed; children are exposed to toxins and environmental dangers from rats to roach killer' (2006: 132). This exposure and unsustainability tie together the social and the ecological to demonstrate the way one affects the other.

Clybourne Park as a name evokes an idea of green lawns, tree-lined streets and safe suburbia. Lena delightfully describes the yard 'with a little patch of dirt where I could maybe get to grow me a few flowers' ([1959] 2009: 70). She is feeling what bell hooks describes as the severed 'humanising connection to nature' (2002: 69) when the Great Migration led Black people to resettle in American cities. One of the motifs of the play is the 'feeble little plant' that Lena tends to determinedly. The plant stays alive despite never having enough sunshine. The plant is the connection to the garden that Lena hopes for. Lena says the plant 'expresses me' ([1959] 2009: 94) as she ensures it will not be harmed in the move. Her children give her surprise gifts of gardening tools and a large ostentatious gardening hat for working in her new yard. The plant is the last thing she takes from the apartment in the final moment of the play. Instead of reading the plant as a metaphor for the American Dream, I argue it is the ecological materiality of the plant that Lena is tied to. For her, the ability to dig in the dirt and garden is part of her liberation. For hooks, re-establishing the severed connection to nature is important now more than ever as 'black folks must collectively renew our relationship to the earth, to our agrarian roots. For when we are forgetful and participate in the destruction and exploitation of the dark earth, we collude with the domination of the earth's dark people, both here and globally' (2002: 70). In the context of American history that explicitly related non-white skin and dirt and therefore 'uncleaniness' and inferiority (Zimring 2016: 4), digging in the dirt and cultivating plants in the yard of her own home is a mode of resistance to the environmental racism of urban structures.

Purchasing the home in Clybourne Park is also an act of resistance for Lena. hooks gives us insight into what life might be like for the family after the move, describing her childhood journey to her grandmother's house in a poor white neighbourhood: 'I remember the fear…because we would have to pass that terrifying whiteness—those white faces on the porches staring us down with hate. Even when empty or vacant, those porches seemed to say "danger," "you do not belong here," "you are not safe"' (1990: 41). She also recalls the feeling of homecoming when she finally entered the yard of her grandmother's house, when her safety was intact again in this 'homeplace'. Lena is building a homeplace in hooks' terms: under white supremacist societies Black people's subjecthood is denied or devalued, but 'black women resisted by making homes where all black people could strive to be subjects, not objects, where we could be affirmed in our minds and hearts despite poverty, hardship, and deprivation, where we could restore to ourselves the dignity denied us on the outside in the public world' (hooks 42). Making a safe place for the affirmation of Black subjecthood, love and community was and is an act of resistance. As in *Raisin* when Black people, particularly women, are working jobs in service of white people, often in domestic situations, a homeplace becomes an important part of a liberation struggle. The political value of the homeplace is made clear through remembrance that denying Black people the ability to make a homeplace (through economic, legal and social structures) has been an effective tool of subjugation (hooks 46). This also makes the homeplace fragile and 'subject to violation and destruction' (hooks 47) such as what violence might face the Youngers in their new neighbourhood. Making and maintaining of homeplaces through mutual care, affirmation and solidarity building is essential to developing a 'meaningful community of resistance' (47). Hansberry's characters recognise the importance of the house in Clybourne Park as a homeplace and decide not to take Mr. Lindner's offer to buy them out, but to move into the home because after five generations of enslaved people and sharecroppers living in the country, they have earned it. The geographies of slavery and postslavery collide with the desire for personhood, manifested in a home of their own with a garden.

While Lena Younger fights to relocate her family and create a homeplace that resists white supremacy structures, she enacts a familiar struggle within environmental justice. Mainstream environmentalism has tended to focus on conservations of 'wilderness' and protections of land, animals and recreation sites, while Rachel Stein asserts that 'the environmental justice movement has instead defined the environment as "where we live, work, play, and worship"' (Stein 2004: 1). In this way, environmental justice brings ecology 'home' and includes 'a range of urban and rural issues that expose poor communities and communities of colour to unfair risks and burdens' (Stein 2). Part of the reason for this is that environmental justice issues tend to be home-related, 'and because women have often been responsible for that domain, women engage in these movements in order to protect and restore the well-being of families and communities threatened by environmental hazards or deprived of natural resources needed to sustain life and culture' (Stein 2). Women often become galvanised in these movements through ideologies of motherhood and family protection, although that extends to community and racial/cultural survival. It is important to note the role sexism plays in determining these gender roles but, as hooks contends, homeplace making as resistance should have a feminist dimension with 'sharing feminist thinking and feminist vision, building solidarity' (1990: 49). In *Raisin*, Lena empowers her son Walter Jr. by trusting him with the remainder of the insurance money (which he eventually loses) and then allowing him to make the decision to move to Clybourne Park or accept Mr. Lindner's buyout offer. Her care for the family certainly conforms to the gender role of the self-sacrificing Black mother but she brings Walter Jr. into her homeplace making, affirming his personhood, which he then models and affirms in his own young son when he refuses Mr. Lindner's offer. In the end, they are united in their commitment to moving to Clybourne Park and building a (fragile) home where they can 'return for renewal and self-recovery, where we can heal our wounds and become whole' (hooks 1990: 49). For the Youngers, home is intersectional: it is a material space where they might escape some of their exposure; it is an ecological space where they can garden as resistance; and it is a space of liberation where after five generations of working under white supremacy, they can affirm and support each other's humanity.

Eco-Cosmopolitanism: A Global Proposal

'If you believe you're a citizen of the world, you're a citizen of nowhere. You don't understand what the very word "citizenship" means' were the words of then UK Prime Minister Theresa May to the Conservative Party Conference in 2016, four months after the referendum vote on Brexit. She was speaking of people 'in positions of power' who acted as if 'they have more in common with international elites than with the people down the road', drawing on some of the same rhetoric used in the Trump US presidential campaign, implying that working people have been cheated by 'international elites', as if May and Trump did not occupy this category themselves. This is an explicit critique of cosmopolitan values that comes at the expense of not just those 'in positions of power' but immigrants, migrants, refugees, those holding multiple citizenships and the many European Union nationals living in the UK. In a shift from left wing politics of the 1990s, globalisation in this rhetoric is often cast as a problem, responsible for the loss of employment or wages of the 'honest, working person'. Globalisation in this sense is often tied to a xenophobic sentiment, placing the blame on immigrants. In the same speech at the Conservative Party Conference, May states that if you are '—and I know a lot of people don't like to admit this—someone who finds themselves out of work or on lower wages because of low-skilled immigration, life simply doesn't seem fair' (May 2016). There is a contradiction in this rhetoric though, as May also states she wants Britain to take a prominent place on the world stage, including in defence of its allies and upholding the Paris Climate Change agreement (although in 2018 the UK Climate Change Committee reported it is not on track to meet their targets (Le Page 2018)). Wanting to play a large role on the world stage while also wanting to close borders and 'control' immigration is a reiteration of colonial imperatives and attitudes as a desire for control and power globally without being held accountable or recognising free movement that is not political or economically advantageous. For Nixon 'hard borders' and walls 'read in terms of neoliberalism and environmental slow violence, materializes temporal as well as spatial denial through a literal concretising of out of sight out of mind' (Nixon 2011: 20). In a global perspective on home, however, citizenship of the world can be a democratising idea that may help us understand the way our choices reverberate in interconnected networks that are global in scale. In ecological terms, we must think of ourselves as 'citizens of the world' so we can understand our intimate relations with humans and more-than-humans across global scales. Instead of 'nowhere' as home, everywhere becomes connected to us because climate catastrophe collapses distance and does not recognise borders. We are intimately connected to the mud of Bangladesh, to the weather being tweeted about across the world, to the political formations that limit where we can live, work and make a home. Instead of absolute deterritorialisation or a sense of nowhere as home, theatre scholar Nadine Holdsworth positions cosmopolitanism as an identity that 'invites us to think of citizens of the world, of global citizens, alongside identities more locally specific to geographical placement or individual subjectivities', although it is also 'about sharing the same planet and recognising the ethical obligations of healthy coexistence that this demands' (2010: 67). Being citizens of the world can mean we understand how those of us living in the industrialised Global North shape and influence the differentiated lives of people living in other places, through our everyday resource-intensive lifestyles, and that this understanding comes with a sense of responsibility to other people and places.

#*CRAZYWEATHER*

It is January in Toronto, which means cold and snowy weather, and slushy streets. Today is bright and sunny although crisp. The previous day was cold with a low of -24° Celsius. Today the temperature gets up to -1 at the height of the sunny afternoon. The temperature is a welcome change after a large ice storm hit Ontario a few weeks earlier, leaving hundreds of thousands of people without power for days. My grandmother had to be coaxed out of her suburban home, not having power or heat but afraid to leave in case the water pipes froze and burst. The destruction caused by the storm is still evident in many parts of the province. Fresh from this recent weathering experience, on this unseasonable January day, I find myself at the

Royal Ontario Museum to see Cape Farewell's *Carbon 14: Climate is Culture* exhibition (2013–2014).

The first piece upon entering the exhibition is comprised of two television screens sitting beside each other on a gallery wall. One screen plays tweets from users around the world, commenting on the weather. The adjacent screen features an animation of the revolving earth as seen from space, NASA's Blue Marble image, on a 10-minute loop. For *#crazyweather* Canadian video-artist Sharon Switzer gathered together tweets from around the world commenting on 'crazy' weather in different localities over the course of a year. Each tweet, such as 'Welcome to Montreal, where the weather is crazy and the seasons don't matter: D' @ aroseagirl, Montreal, Quebec, 9/19/2013 2:47:08 PM', is a Twitter user commenting on the rapidly changing or unseasonable or extreme weather in their daily lives. The tweets include snow in May and September, heat in April and extreme events like hail, flooding, storms, monsoons, Hurricane Sandy and 'sideways' rain. They also highlight extreme changes in weather such as 'Trying to figure out how to convert the boat I needed yesterday into a sled for today #crazyweather @ LukasBolt, Independence, Virginia, 01/31/2013 4:19:38 PM'. Twitter is only one of the intensified networks of global exchange that permeate our everyday lives. The tweets fly in and out on one screen, with the name, location, date and time of each noted below them, while a projection of the Earth as seen from space slowly rotates beside them. As I stare at the screens, mesmerised, the installation slowly builds a picture of the extent, volume and reach of these 'crazy' weather events and a feeling of dread begins to creep in. Understanding these events as a global pattern puts localised weather experiences into a wider context and allows us to see the interrelated mesh of relations in which our everyday lives are situated in. The work is also an effective visualisation of a global sense of place or home.

#crazyweather registers an eco-cosmopolitan understanding of relations to place and home. Eco-cosmopolitanism, as elaborated by Ursula K Heise (2008), acknowledges that we all share the earth in which we live, rather than employing cosmopolitanism as a synonym of mass culture or a global idea of governance. Cosmopolitanism comes from the Greek *kosmos* meaning world and *polis* meaning city. However, the idea of a world city or one-world culture is problematic in many different respects. As cultural theorist Timothy Brennan (1997) points out, it seems that a 'world community' actually meant an adoption of 'American-ness' across the world, which replaced colonialism with globalisation (in a neocolonial fashion),

and which maintained the same inequalities through 'the reliance of the overdeveloped world on cheap labour, higher rates of exploitation, and extraction of raw materials' (Brennan 5). In *Sense of Place and Sense of Planet* (2008), Heise describes contemporary cosmopolitanism as a heightened global interconnectedness (57) or 'eco-cosmopolitanism' which 'reaches towards…the "more-than-human world"—the realm of nonhuman species, but also that of connectedness with both animate and inanimate networks of influence and exchange' (Heise 60–61). My interpretation of cosmopolitanism is not a totalising system of capital exchange. Rather, as iterated through eco-cosmopolitanism, it upholds and acknowledges difference and seeks to underscore the way in which we are implicated in ecological global networks, through everyday activities. I propose a qualification to Heise's conception though; rather than connectedness to animate and inanimate networks of exchange, I assert that all networks are animate in one way or another, in an ecomaterialist way. Twitter, in *#crazyweather*, is an animate network, which not only spreads information and advertising, connects people and influences politics, it also has an ecological 'footprint'. Tweets are powered by data centres (or server farms) which use a large amount of energy. The emissions produced by data centres for digital services are responsible for 2% of global CO_2 emissions (Vaughan 2015). The effects produced by a tweet make it an animate and agentic network that produces material ecological effects.

The duality between local and global is also a duality of difference and similarity, between humans themselves as well as between humans and other species and 'things' (Heise 2008). Negotiating these dualities is inherent in ecological art practice as well as the critical theories used to interrogate them. The effects of extreme weather on more-than-human species is acknowledged in a tweet by @Shardulbajikar from Mumbai, India, 'Glad summer's now warming up. Crazy weather patterns r not good for plants, insects, birds & yes humans too! 4/24/2013 9:21:47 pm'. Our experience of the more-than-human world is often local and situated, which makes it difficult to comprehend how we are firmly bound up in global ecological networks of capital, information, pollution, consumer items, food, travel, labour, waste and weather.

#crazyweather takes a well-known image of the planet (NASA's Blue Marble video) and juxtaposes it with new heterogeneous informational networks made up of different global users. Through the combination of different scales that reflect 'connectedness to both informational and social networks that span the world', the work 'suggests some of the

complexities an eco-cosmopolitan imagination of the global must take into account' (Heise 2008: 67). Heise argues that this global sense of place requires deterritorialisation, or the severing of social and political practices from places or specific cultures, as the increasingly connected globalised world necessitates the development of 'new forms of culture that are no longer anchored in place' (2008: 10). Twitter is one new form that is not rooted in place. *#crazyweather,* through the deterritorialisation of local knowledge and weather events, 'opens up new avenues into ecological consciousness' (Heise 162) towards a global ecological understanding. For Heise, the value of cosmopolitan discourses to an ecological project is the way in which they may facilitate thinking beyond the individual self, outside the limitations of a bounded culture, race or nation, to broader sociocultural understandings and ways of thinking (Heise 60). *#crazyweather* takes individual users in different places and makes them part of a broader sense of climate, extending beyond the individual and seemingly isolated weather events to draw a picture of global weather patterns. The tweets connect people in places such as Vancouver, Canada; Islamabad, Pakistan; Riyadh, Saudi Arabia; Om El Donya, Egypt; and Helsinki, Finland. Heise conceives cosmopolitanism as a planetary 'imagined community' (following Anderson)[4] and explores the way in which cultural works address the perceptual ties and modes of identification with the natural world. An eco-cosmopolitan analysis, however, requires attentiveness to the current political and social conditions of specific communities and differentiated ecological relationship, which may diminish or problematise how people may be able to envision themselves as part of a 'planetary community'. Heise's claim of deterritorialisation can ignore the material realities of specific localities (the worsening effects of climate change on the Global South, for example). For those that do not have access to Twitter, or a reliable internet connection or a computer, *#crazyweather* is not something they can participate in. They are not represented as part of this 'imagined planetary community'. These may also be the people most affected by 'crazy' weather and other climate change effects. Recognising themselves as part of a planetary community would be difficult and may reinforce colonial and neocolonial hierarchies for those without access to technological networks, political capital or global mobility. As the differentiated effects of climate change are disproportionately felt, imagining this work as inclusive as a planetary community elides these differences and homogenises, rather than recognising the asymmetrical violence of 'crazy' weather.

The piece references the weather not as a fixed or continual 'object', rather as processual, unpredictable and unstable. Referring to the weather as dramaturgy, David Williams describes it as 'dynamic, non-homogeneous environments forever in process, mutable, an ever-changing flux of micro-events in a relational field of infinite complexity…' (2006: 142). The microevents of weather, documented on Twitter and accumulated in *#crazyweather,* allows for 'considering ecological crisis and environmental as well as cultural connectedness across different scales' (Heise 2008: 209) and new ways of imagining and visualising the earth as home. To a certain extent, our lived experience of weather will happen on a local scale. As Williams suggests 'weather is always contextual, and at the same time in the ephemeral spatio-temporal events that characterize its happenings the local is invariably implicated in the trans-local' (Williams 2006: 142). Weather is felt and experienced locally but it reaches beyond locality to something bigger, atmospherically and culturally. As I argued in Chap. 3, following Neimanis and Hamilton (2018), the weather means more than 'temperature, wind speed/direction, humidity, pressure, atmosphere, resultant phenomena of various kinds' (Williams 2006: 142), it is social forces, political conditions, colonialist imperatives, histories, cultural representations, sensory experiences, ecological exposure—in other words, it is us.

An eco-cosmopolitan reading posits the performance as considering the way in which the more-than-human (specifically the weather) is imbricated in relationships with the human. The seemingly innocuous tweets take on a different meaning when paired with the image of the earth and the sheer number and variety of the 'crazy' weather occurrences. A larger picture and context of the global effects of climate change begin to emerge. The exhibition brochure notes that 'while the individual comments may seem familiar and anecdotal, even unimportant, their cumulative effect is rather troubling' (Cape Farewell 2013). There are high winds in Sydney, Australia and Ulaanbaatar, Mongolia, each noted by separate individuals. The connectedness of these weather events, through virtual space and through the material conditions of their environments, creates a sense of unsettlement, almost alarm at the sheer extremity of the weather. The crazy weather accumulates, giving the viewer a sense of the slow violence of climate change (Nixon 2011). For artist Switzer, the volume of crazy weather is 'signalling some serious things going on with climate change and global warming' (Cape Farewell 2013). This sense of alarm and the banality of the tweets (about the weather, the most ubiquitous conversational material) enact some of the tension between the local and global effects of climate change evident in our everyday lives and creates a compelling narrative of global environmental catastrophe.

Carla and Lewis

Elsa, a New York-based, white, middle-class visual arts curator, is developing an interactive installation called The Amina Project, to draw attention to the effects of climate change on the people in Bangladesh. Elsa is inspired after reading an article of a woman named Amina who is in distress over climate change effects in Bangladesh. For the art piece New York-based gallery goers will be able to Skype with Bangladeshi climate change refugees in order to hear their stories. Elsa's background and privilege allows her to ignore the objectification of climate change refugees and the urgent ecological crisis in her own home. She commissions artists Carla and Lewis to create work in response to the Skype conversations in the gallery. Carla and Lewis are punk butterflies, queer ecological creatures who are called 'little girls' by the other characters but blur the gender and species lines. Carla and Lewis camp out in Elsa's apartment to develop their work for the installation. The apartment is slowly being taken over by mud: there is a thick layer on the floor, and it comes through the walls until eventually it takes over. The mud is performed by an ensemble of actors, portraying an assemblage of human and more-than-human actants including landscape, drawings, subway cars: *'out of the mud come: Crocodiles. Malaria. Rotting wood'* (Enelow 2014: 88). This is the story presented in the play *Carla and Lewis* (2014).

Carla and Lewis was developed as part of the Ecocide Project, a research theatre project to investigate how climate change could be represented in theatre in all of its complexity. Written by Shonni Enelow, directed by Fritz Ertl and Josh Hoglund with Una Chaudhuri as dramaturg, it premiered in 2011 at the Incubator Arts Project in New York City. The play's structure, characters and performance style enact the connections between local and global ecological relationships while highlighting the asymmetrical effects of climate change and the ability to respond, particularly for postcolonial/neocolonial subjects in the Global South.

In the play, curator Elsa recruits Bangladeshi feminist mixed-media artist Kamna Benerjee to work on the project from Bangladesh. Elsa asks her to find 'Amina', her village, her shack—or a woman in an analogous situation. She sends her the *New York Times* article in which she learned about

Amina. Kamna responds that this will be impossible, she lives in the city of Dhaka nowhere near Amina's southern village, and besides, it will be unnecessary. 'Amina' cannot be the face of climate change because climate change is too diffuse and decentralised: 'It's as if this very structure has closed it off from representation' (Enelow 2014: 95). She goes on to say that 'Amina' as a symbol of climate change does not make sense because 'climate change is enormous, it's tiny, it's impossible, it's happening—all the way up and all the way down' (96). The play suggests that no human can be the face of climate change (especially a naive concept of a postcolonial Bangladeshi villager). Rather, in the play, punk butterflies Carla and Lewis represent climate change, as well as the relations between everything else. The queer creatures that blur species lines and operate in a no future ideology are a more complex symbol of the messiness of climate change.

The mud is also a central figure in the play, slowly invading Elsa's apartment and providing a landscape that underscores the ecomateriality of the play. The stage directions such as '*The weather changes. Rotting wood. Dead fish. Computer parts. Amina.*' gesture to the relationality and messiness of climate change, infecting and infected by everything else. The mud landscape is played by a company of actors, who also perform the stage directions. Shifting and changing, they represent the human and more-than-human assemblages and relationships, enacting a kind of bioperformativity. Aron and May write that the landscape is 'presented as both a physical reality and as a landscape of the artists' creation, the play's ecosystem blurs distinctions between reptiles, mammals, viruses, plants, objects and humans, democratizing and horizontalizing their status' (Arons and May 2013: 192). The actors playing the mud and stage directions are not only representative however, they are also material. During the play, performers referred to as Actor 1 and Actor 2 address the audience:

> Actor 1: This is the scene in which we introduce the world and its landscape, which is of course this theatre: its flora and fauna, its technological machinations, and its natural and man-made ornamentation...there are approximately 80 metal chairs. The metal is aluminum, the silvery white member of the boron group of chemical elements. It is the most common element in the earth's crust. Plants ingest it in their food, the soil, as do animals, who ingest plants. You ingest it.
> Actor 2: You put it under your arms if you use antiperspirant.
> (Enelow 2014: 91)

This direct address facilitates an ecomaterialist understanding of how the aluminium seating and the human audience both have the same material in them (albeit in different concentrations). The flora and fauna of the theatre is spotlighted and considered as the risers, the seating, the technical equipment, the ornamentation and the audience. Chaudhuri has previously written about the deeply humanist tradition of theatre that tends to reinforce the nature/culture divide rather than re-imagine it (1994). Addressing the theatre's flora and fauna reconfigures that divide and seeks to reveal the relationships and affinity between a human audience member and the seat they are sitting in. The Actors also ask the audience how they travelled to the theatre, whether they took the subway and what the experience of that was. These rhetorical questions reframe the theatre-going experience as ecological, drawing the material connections between humans and things with the capacity of agency.

The mud is also agentic, as Chaudhuri and Enelow write, as it was conceived to bring Jane Bennett's 'thing-power' into the theatre (2014: 34). 'Mud was more than a material presence: it was a non-human force, an "actant"… It drove the story and affected the characters in both obvious and unexpected ways. But it was also an assemblage' of lively material (Chaudhuri and Enelow 34). It is this dramaturgical gesture of connections and human/more-than-human assemblages that seek to make climate change meaningful to a New York audience through understanding their material interactions and vibrant relationships to agentic things that extend across place. It shares some of the aims of ecomaterialism: a sense that we are connected in differentiated ways, through networks of exchange, to the more-than-human. From an eco-cosmopolitan perspective, Heise describes the global networks of exchange in our daily lives:

> From the food, clothes, and fuel we buy to the music and films we enjoy, the employer we work for, and the health risks we are exposed to, everyday routines for most people today are inconceivable without global networks of information and exchange. (Heise 2008: 54)

These everyday things are also all ecological, creating effects across these networks. The performed assemblage of 'mud' (made of actors who also play *Rotting wood. Dead fish. Computer parts. Amina*) as the landscape of the play creates an eco-cosmopolitanism dramaturgy and demonstrates the way in which ecological relationships are global in scale, implicating a global sense of place and a more-than-human community.

Eco-cosmopolitanism is a way of looking beyond boundaries of animate/inanimate, human/animal, or human/nature, to the ecological assemblages of exchange that encompass the more-than-human. The mud is a more-than-human material that invades Elsa's New York apartment ignoring borders and boundaries of home as it is the mud of the Bangladesh flooding that displaced 'Amina'. Chaudhuri writes in the programme notes of the play that 'the mud of Bangladesh takes revenge, and its action is no longer operating at a distance. Just as contemporary artists seek actual encounters and meaningful proximities, just as the forces of global capital compel transnational flows of images, things, and substances, climate change takes its revenge' (Chaudhuri in Enelow 2014: 123). The mud slides of Bangladesh infect the New York apartment, paying no attention to the geographical distance or borders between them, disrupting ideas of place and home as bounded and delimited. Arons and May suggest that 'through this substitution of the hyperlocal for the global, the play makes clear how illogical and ineffectual our response to climate change and its effects has been' (Arons and May 2013: 192). The slow violence of the mud mirrors the slow violence of climate change. It does not inhabit an 'over there' distant place that can be ignored or dismissed as 'primitive' or 'underdeveloped' and therefore less consequential in neocolonial logic. The mud is enacting an eco-cosmopolitanism, one that does not ignore the differences between people and their mobility, but that is premised on a sense of planet as place and as Heise argues, 'highlight[s] the imbrication of local places, ecologies, and cultural practices in global networks' (Heise 2008: 210). The mud is cumulative and is also more than material mud. It enacts abstract connections to highlight the imbrication of material effects and practices.

Eco-cosmopolitanism shares similarities with what Vandana Shiva (2005) calls 'Earth Democracy', which 'addresses the global in our everyday lives, our everyday realities, and creates change globally by making change locally' (Shiva 4). Her concept of earth democracy suggests a particular local/global relationship and acknowledges, like Heise, the way in which our everyday lives are part of global networks with far-reaching ecological effects. Shiva also extends the idea of community to include the more-than-human: 'Earth Democracy connects the particular to the universal, the diverse to the common, and the local to the global. It incorporates what in India we refer to as *vasudhaiva kutumbkam* (the earth family)—the community of all beings supported by the earth' (Shiva 1). Both earth democracy and eco-cosmopolitanism call upon us to recognise our part within a global yet differentiated community of the more-than-human, involved in interconnected exchanges across multiple scales of time and place, and the adverse and violent effects of these relationships.

Eco-cosmopolitanism must not substitute understanding of difference for a homogenous sense of a 'global planetary community'. Our ecological relationships are bound up with social and political power formations, with ethical dimensions operating on a globe-spanning scale. Elsa's configuration of the Bangladeshi climate change refugee, 'Amina', is premised on colonial ideology of a marginalised Other in the Global South. Kamna writes that she does not have time to go to the south and find 'Amina' because she is getting married and moving house, besides she does not think Amina 'will have anything interesting to say. She was poor before and she is still poor. She was wretched before and she is still wretched' (Enelow 2014: 96). Elsa responds in her email, not to the ideas Kamna presents, but to the news of her marriage and her moving house. 'What a coincidence about your housing stress—I've been having tremendous apartment stress too …moving is such a terrible hassle, isn't it?' (96). The irony of the having to move home (from mud leaking into her New York apartment) is lost on Elsa. She does not see her privileged mobility. Elsa finally responds that 'Amina' is essential to the project as the human face of climate change: 'If everything is de-centred, if climate change is so far beyond ourselves, there's no room for moral responsibility, is there? But there *is* moral responsibility. There *is* Amina' (106). Kamna cannot agree with her conclusion, questioning the whole premise of the project: Skype conversations in a gallery with Bangladeshi climate refugees. She fears this 'would be awkward and superficial, if not nonsensical' and chastises Elsa for fetishising 'the "real" Bangladeshi peasant' (110). Elsa needs a human representation of climate change but Kamna resists this flattening of difference and easy sense of 'connection' to a construction of the 'other'.

Kamna is adamant that there is no human face of climate change, rather 'if there is a face of climate change, it's the fender of the American SUV' (Enelow 2014: 110). The reference to the American SUV may be citing a form of environmental racism that results from the colonial and neocolonial power structures that imposed Western economic models of production for export, as Westra and Wenz suggest 'cultural imperialism is thus a form of cultural pollution that leads to violence, environmental pollution and death' (1995: xxii). Not only does Elsa fetishise 'Amina' as a Bangladeshi peasant, she extends this colonial thinking to Kamna, denying her a sponsorship letter so that she might get a visa to travel to New York and discuss the art piece as a dialogue between artists. This neocolonial logic is evident in America's attitude to the Global South on environmental issues, which Enelow describes as 'don't do as we do, don't consume like we consume' (in Hoglund et al. 2014: 77) while still being considered 'ground zero' for irresponsible environmental behaviour. In this gesture, Elsa is demonstrating some of the tension in the idea of cosmopolitanism, which erases the alterity of marginalised communities within the idea of a globalised world. 'HUMANS! LEAVE YOUR HOMES' is a repeated phrase in the play but not all humans have a place to flee to. Amina, for example, has nowhere to go as a climate change refugee. As Arons and May assert, having nowhere to go is 'a dilemma faced by climate refugees of all species' (2013: 193). This includes Kamna, 'Amina' or indeed many other species beyond the human do not have the same 'global mobility' as Elsa. The borderless world that Elsa imagines, in which climate change effects ignore national boundaries, and international dialogues in galleries help Americans understand their contribution to climate change, is not the world that Kamna and Amina inhabit.

Kamna's emails demonstrate that migrants, refugees and nomads do not have the same sense of a global world that some theories of cosmopolitanism seem to suggest, because they do not have access to global mobility. The idea of home and place may be contested for these peoples and their 'circulation' is often politically, culturally or socially motivated. Cosmopolitanism that ignores the mobility (or lack of mobility) of marginalised peoples may render it an elitist or ethnocentric idea. The status of the subaltern classes (groups of peoples who are outside of the hegemonic, often colonialist power formations, socially, politically and/or geographically), within global capitalism, complicates the cosmopolitan dream of global mobility in a world without recognised borders. Elsa will not be able to produce an artwork that 'communicate[s] across cultures—so that Americans can understand that this is not some abstract phenomenon' (Enelow 2014: 106) without understanding and addressing the colonial and neocolonial structures that produced the 'Bangladeshi peasant' and the 'American SUV'.

Elsa's frustration in the mud infecting her home mirrors the frustration of Kamna in not having her voice heard in the project and can be read as the frustration of all those marginalised people who are bearing the brunt of the burden of climate change effects while contributing very little to its onset. Elsa calls and calls but the building management ignore her mud leak. She yells but is ignored. When Carla and Lewis confront her about the mud and sink hole her home is becoming she responds: 'I KNOW, I

KNOW IT'S EVERYWHERE, I keep calling and calling and NOBODY ANSWERS ME!…I have been screaming, how can I scream louder, they KNOW, they KNOW what's going on, and they just don't care' (Enelow 2014: 113). Similarly, people living the Global North understand that climate change is happening, but most Western governments pursue very little action on it, instead favouring short-term political gain (i.e. building oil and gas pipelines, fracking, etc.). Elsa's frustration at the Bangladeshi mud is the same frustration of all of the people who are disproportionately affected by the emissions-heavy lifestyles of the Global North. Often it is Indigenous peoples, without much social and political capital, who are 'calling and calling' for something to be done. Elsa does not recognise her responsibility or response-ability in the mud. She wants someone else to handle it, rather than attempting to fix it herself, which mirrors the presumptions of the Global North in addressing climate change. Although an individual cannot fix the eco-political situation, there are levels of responsibility and response-ability that Elsa has as a privileged citizen. Rather than taking responsibility, waste is still shipped to other countries; blame is shifted, waiting for someone else or the 'markets' of neoliberalism to take care of it.

Carla and Lewis enacts the complexity, frustration and presumptive ideologies of climate change, drawing eco-cosmopolitan connections across people and places while recognising the differentiated way in which humans and more-than-humans are implicated and marginalised.

Places of Difference

Ecodramaturgies across different forms of theatre and performance respond to and reflect current local and global ecological conditions and understandings of home, informing modes of practice, presentation and reception. Ecodramaturgies offer aesthetic provocation for forms, images, stories and experiences of eco-cosmopolitanism, while also resonating with local socio-political-ecological systems. It can open up a way of viewing work and the wider world. Negotiating the complex relationship between a globalised world and local ecological conditions and understanding a planetary sense of home without ignoring differences, ecodramaturgies can reveal and interrogate some of the nuances of this complexity.

Trans-Plantable Living Room and *Above Me the Wide Blue Sky* enacted ideas of home as local, gesturing and invoking a romantic nostalgia while inferring the sense of loss and ecological material of the earth as home.

Home is also shaped by racialised power formations that delimit access to home in an ecological sense, and is often riddled with environmental racism. Homeplace making and gardening become an act of resistance in *A Raisin in the Sun*, paralleling the same impulse that has led to environmental justice work by women of colour. *#crazyweather* and *Carla and Lewis* manifest some of the ways in which local networks, practices and experiences are actually part of global ecological exchanges, revealing the way everyday habits and neocolonial logics have far-reaching effects, shaping concepts of place. All of these distinctive works refract global ecological relationships by elucidating different aspects and layers of global, more-than-human assemblages.

Notes

1. Some of the material in the following section was drawn on for a short co-authored paper with Bronwyn Preece, 'Trans-Plantable Living Room', published in *Contemporary Theatre Review* (2015) and a co-authored (with Preece and Meghan Moe Beitiks) article in the *Centre for Sustainable Practice in the Arts Quarterly* magazine (2015).
2. For more on ecoscenography, see Beer, Tanja. 2021. *Ecoscenography: an introduction of ecological design for performance*. London; New York: Palgrave Macmillan.
3. See Woynarski, Lisa. 2020. Towards Radical Coexistence in the City: Performing the bio-urban in Bonnie Ora Sherk's *The Farm* and Mierle Laderman Ukeles's *Flow City*. *Performance Research* 25:2: 124–131.
4. Anderson's ([1983] 2006) concept of imagined communities is one of a socially constructed community, 'imagined' by those who identify themselves as part of it. For him, the concept of nation was an imagined community of people who do not actually know each other but perceive themselves to be part of the identifying, shared community.

References

Anderson, Benedict R. O.' G. [1983] 2006. *Imagined Communities: Reflections on the Origin and Spread of Nationalism*. Revised ed. London; New York: Verso.

Arons, Wendy, and Theresa J. May. 2013. Ecodramaturgy in/and Contemporary Women's Playwrighting. In *Contemporary Women Playwrights: Into the Twenty-First Century*, ed. Penny Farfan and Lesley Ferris, 181–198. London; New York: Palgrave Macmillan.

Beer, Tanja. n.d. What Is Ecoscenography? Accessed May 18, 2019. https://ecoscenography.com/what-is-ecoscenography/.
Bennett, Jane. 2010. *Vibrant Matter: A Political Ecology of Things*. Durham: Duke University Press.
Blunt, Alison, and Robyn M. Dowling. 2006. *Home*, Key Ideas in Geography. New York: Routledge.
Brennan, Timothy. 1997. *At Home in the World: Cosmopolitanism Now*, Convergences. Cambridge: Harvard University Press.
Cape Farewell. 2013. *#crazyweather*. Cape Farewell. Accessed February 04, 2019. https://www.capefarewellfoundation.com/carbon14/crazyweather/.
Chaudhuri, Una. 1994. "There Must Be a Lot of Fish in That Lake": Toward an Ecological Theater. *Theater* 25 (1): 23–31.
———. 1995. *Staging Place: The Geography of Modern Drama*. Ann Arbor: University of Michigan Press.
Chavis, Benjamin F., Jr. 1992. Foreword. In *Confronting Environmental Racism: Voices From the Grassroots*, ed. Robert D. Bullard, 3–6. Boston: South End Press.
Cieraad, Irene. 1999. *At Home: An Anthropology of Domestic Space*. Syracuse: Syracuse University Press.
Cloke, Paul, Paul Milbourne, and Rebekah Widdowfield. 2000. Homelessness and Rurality: 'Out-of-Place' in Purified Space? *Environment and Planning D: Society and Space* 18 (6): 715–735.
Cresswell, Tim. 2004. *Place: A Short Introduction*, Short Introductions to Geography. Malden, MA: Blackwell Publishing.
Enelow, Shonni. 2014. Carla and Lewis. In *Research Theatre, Climate Change, and the Ecocide Project: A Casebook*, ed. Una Chaudhuri and Shonni Enelow, 87–116. London; New York: Palgrave Macmillan.
Fevered Sleep. 2013. *Above Me the Wide Blue Sky*. Performance, Young Vic, London, March 27.
Hansberry, Lorraine. [1959] 2009. *A Raisin in the Sun*. Reprinted. Methuen Drama Modern Classics. London: Methuen.
Harré, Rom, Jens Brockmeier, and Peter Mulhausler. 1998. *Greenspeak: A Study of Environmental Discourse*. Los Angeles; London: Sage Publications.
Harvey, David. 1996. *Justice, Nature, and the Geography of Difference*. Malden, MA: Blackwell Publishing.
Heise, Ursula K. 2008. *Sense of Place and Sense of Planet: The Environmental Imagination of the Global*. Oxford; New York: Oxford University Press.
Hoglund, Josh, Sunita Prasad, Meng Ai, and Shonni Enelow. 2014. Staging Carla and Lewis. In *Research Theatre, Climate Change, and the Ecocide Project: A Casebook*, ed. Una Chaudhuri and Shonni Enelow, 62–86. London; New York: Palgrave Macmillan.
Holdsworth, Nadine. 2010. *Theatre & Nation*. London; New York: Palgrave Macmillan.

hooks, bell. 1990. *Yearning: Race, Gender, and Cultural Politics*. Boston: South End Press.

———. 2002. Earthbound. In *The Colors of Nature: Culture, Identity, and the Natural World*, ed. Alison Hawthorne Deming and Lauret E. Savoy, 67–71. Minneapolis: Milkweed Editions.

Huggan, Graham, and Helen Tiffin, eds. 2010. *Postcolonial Ecocriticism: Literature, Animals, Environment. London*. New York: Routledge.

Lavery, Carl, and Simon Whitehead. 2012. Bringing It All Back Home: Towards an Ecology of Place. *Performance Research* 17 (4): 111–119.

Lawson, Bill. 1995. Living for the City: Urban United States and Environmental Justice. In *Faces of Environmental Racism: Confronting Issues of Global Justice: Confronting Issues of Global Injustice*, ed. Laura Westra and Peter S. Wenz, 41–55. Lanham, Maryland: Rowman & Littlefield.

Le Page, Michael. 2018. UK Is Not on Track to Meet Its Own Climate Targets, Says Report. *New Scientist*, June 28. https://www.newscientist.com/article/2172829-uk-is-not-on-track-to-meet-its-own-climate-targets-says-report/.

Mackey, Sally. 2007. Performance, Place and Allotments: Feast or Famine? *Contemporary Theatre Review* 17 (2): 181–191.

Mackey, Sally, and Nicolas Whybrow. 2007. Taking Place: Some Reflections on Site, Performance and Community. *Research in Drama Education: The Journal of Applied Theatre and Performance* 12 (1): 1–14.

Massey, Doreen B. 1997. A Global Sense of Place. In *Reading Human Geography: The Poetics and Politics of Inquiry*, ed. Trevor J. Barnes and Derek Gregory, 315–323. London; New York: Arnold; Wiley.

———. 2007. *World City*. Cambridge, UK; Malden, MA: Polity.

Matthews, Kristin L. 2008. The Politics of 'Home' in Lorraine Hansberry's A Raisin in the Sun. *Modern Drama* 51 (4): 556–578.

May, Theresa J. 2006. 'Consequences Unforeseen…' in Raisin in the Sun and Caroline, or Change. *Journal of Dramatic Theory and Criticism* 20 (2): 127–144.

May, Theresa. 2016. Theresa May's Keynote Speech at Tory Conference in Full | The Independent. *Independent*, sec. UK Politics, May 10. https://www.independent.co.uk/news/uk/politics/theresa-may-speech-tory-conference-2016-in-full-transcript-a7346171.html.

McKittrick, Katherine. 2013. Plantation Futures. *Small Axe* 17 (3): 1–15.

Meyer, Stephen Grant. 2001. *As Long as They Don't Move Next Door: Segregation and Racial Conflict in American Neighborhoods*. Lanham, Maryland: Rowman & Littlefield.

Neimanis, Astrida, and Jennifer Mae Hamilton. 2018. Weathering. *Feminist Review* 118 (1): 80–84.

Nixon, Rob. 2011. *Slow Violence and the Environmentalism of the Poor*. Cambridge: Harvard University Press.
Preece, Bronwyn, Meghan Moe Beitiks, and Lisa Woynarski. 2015. The Trans-Plantable Living Room: Sites, Processes and Performances. *CSPA Quarterly* (12): 18–25.
Pulido, Laura. 2000. Rethinking Environmental Racism: White Privilege and Urban Development in Southern California. *Annals of the Association of American Geographers* 90 (1): 12–40.
Rose, Gillian. 1993. *Feminism and Geography the Limits of Geographical Knowledge*. Cambridge, MA: Polity.
Saunders, Peter, and Peter Williams. 1988. The Constitution of the Home: Towards a Research Agenda. *Housing Studies* 3 (2): 81–93.
Scruton, Roger. 2012. *Green Philosophy: How to Think Seriously about the Planet*. London: Atlantic.
Seamon, David, and Robert Mugerauer, eds. 1985. *Dwelling, Place, and Environment: Towards a Phenomenology of Person and World*. Dordrecht, Netherlands: Martinus Nijhoff Publishers.
Shiva, Vandana. 2005. *Earth Democracy: Justice, Sustainability, and Peace*. Boston: South End Press.
Stein, Rachel. 2004. *New Perspectives on Environmental Justice: Gender, Sexuality, and Activism*. New Brunswick; London: Rutgers University Press.
Tuan, Yi-Fu. 1974. *Topophilia: A Study of Environmental Perception, Attitudes, and Values*. Englewood Cliffs, NJ: Prentice-Hall.
Vaughan, Adam. 2015. How Viral Cat Videos Are Warming the Planet. *The Guardian*, September 25. https://www.theguardian.com/environment/2015/sep/25/server-data-centre-emissions-air-travel-web-google-facebook-greenhouse-gas.
Watts, Stephen. 2013. Interview by Rosie Leach. Trans-Plantable Research. Cardiff, Wales, May 23.
Westra, Laura, and Peter S. Wenz, eds. 1995. *Faces of Environmental Racism: Confronting Issues of Global Justice: Confronting Issues of Global Injustice*. Lanham, Maryland: Rowman & Littlefield Publishers.
Williams, David. 2006. Weather. *Performance Research* 11 (3): 142–144.
Woynarski, Lisa, and Bronwyn Preece. 2015. The Trans-Plantable Living Room. *Contemporary Theatre Review* 25 (3): 421–423.
Zimring, Carl A. 2016. *Clean and White: A History of Environmental Racism in the United States*. New York: NYU Press.

CHAPTER 6

Decolonised Ecologies: Performance Against the Anthropocene

Banners with Indigenous artwork saying '*Mni Wiconi*: Water is life'; lit up letters spelling out 'Protectors' and 'Protect the Sacred'; riders on horseback; energised young people with painted faces and Neil Young—these are just a few of the images from Standing Rock circulating on social media. The artwork and images from the NoDAPL (No Dakota Access Pipeline) camps helped to generate media coverage and international awareness of the infringement of treaty rights by the building of the Dakota Access oil pipeline. From April 2016 to February 2017, Indigenous communities, activists, artists and allies camped out—acting as water protectors, with the Standing Rock Sioux tribe—to oppose the development of the Dakota Access underground oil pipeline in North and South Dakota, US. The pipeline cuts through the water sources and burial sites of the Standing Rock reservation, which forbids pipelines on sacred land, threatening their land and water. The pipeline crosses the Cannonball River, on the ancestral lands of the Lakota Sioux. The history of pipelines leaking makes spills and breaks likely,[1] threatening the Missouri River: both a source of livelihood and a significant cultural connector for the Standing Rock Sioux. In the face of violent reactions by multi-state law enforcement and the National Guard—including rubber bullets, sound and water cannons and attack dogs—many Indigenous Nations and peoples (from across the Americas) came together in solidarity. As a result, the histories of settler colonial violence towards Indigenous peoples and land were, at least, *momentarily* remembered.

L. Woynarski, *Ecodramaturgies*, New Dramaturgies,
https://doi.org/10.1007/978-3-030-55853-6_6

Mni Wiconi ('water is life') is a Lakota phrase that became synonymous with the movement and, as Jessica Horton writes, 'augmented Indigenous resistance to state and corporate abuse by communicating the noninstrumental value of nurturing liquid connecting human bodies, other beings, and the land' (Horton 2017: 68). The Lakota have a relationship to water that is not solely based on resourcism; it is cultural and spiritual. Chief Arvol Looking Horse, the 19th Keeper of the Sacred Bundle and Spiritual Leader of the Lakota, Dakota, Nakota People, articulates water of life as 'our first medicine' and 'part of our creation story, and the same story that exists in many creation stories around Mother Earth' (2018). Not only is access to clean water a human right, essential to survival, it is part of the creation stories of many Indigenous peoples, central to their cosmologies. Through the art and stories shared through social media, the significance of water as so much more than a resource countered the colonial, capitalist narrative. Even though the pipeline was built and is currently in operation, Looking Horse does not see the Standing Rock protests as a failure, but the beginning of an international movement: 'it was at Standing Rock that so many came together to share their stories and knowledge of what was happening in their territory, sharing ideas on how to move toward sustainable living in our relationship to land, water, and food' (Looking Horse 2018). The critique of colonial and neocolonial environmentalism, global capitalism and extractivism was brought to a global audience through the movement and art produced at the Sacred Stone and Oceti Sakowin Camps. The water protectors and the NoDAPL movement demonstrate that it is not 'humans' as a homogenised group that are responsible for the epoch of the Anthropocene, but structural and state violence based on ideologies of white supremacy, colonialism and resourcism.

In this chapter, I focus on a mix of activist performances, protests and play texts taking place in North America.[2] The protests and performances look at the stories that shape understandings of our relationship to both the human and the more-than-human world. Stories are powerful forms of representation in many Indigenous cultures, and performance is one of the central places where these stories are shared, celebrated and enacted. Performance is a powerful site of cultural imagery for Indigenous ecological knowledges. I use the terms 'Indigenous' and 'Indigenous peoples' in their global usage, although specific Tribes, Nations and cultures are named where available. Indigenous[3] is capitalised to refer to the identity and particular peoples, rather than its usage as an adjective, to follow Michael Yellow Bird's call 'to signify the cultural heterogeneity and political sovereignty of these groups'

(Bird 1999: 2). 'Indigenous peoples' is used to acknowledge a shared history of colonial violence that is not homogenous, nor to refer to a racial monolith. According to the Center for World Indigenous Studies Fourth World Atlas Project from 2002, there are 1.3 billion Indigenous peoples globally, with over 6000 nations (Rÿser 2015), 'distinguishable according to language, behavior, dress, geography, foods, technologies, creation stories, and numerous other characteristics' (Bird 1999: 3).

My central argument here is that Indigenous ecodramaturgies are needed to decolonise the Anthropocene discourse and the underlying idea of the human/more-than-human relationship it implies. Indigenous ecodramaturgies are what Dione Joseph refers to as 'practices [that] contextualize the political, social, and spiritual development of Indigenous narratives. Subsequently, these are brought together under an overarching framework that allows specificity of individual practices and spaces to be valued and legitimized on their own terms' (Joseph 2019: 132). Such ecodramaturgies manifest a relationship to the past, present and sense of future that is located in Indigenous ecological thinking, and they resist the Anthropocene narrative of the singular 'human' as responsible for the earth's destruction. The NoDAPL and Idle No More activist movements enacted an Indigenised form of environmental justice, bringing into relief the federal laws and structures that ignore Indigenous relationships to land and ensure domination over First Peoples through ecocide. Staged performances and rituals were used in these movements to reframe the story of domination to sovereignty, land rights and healing.

Turning to scripted plays, *Burning Vision* (2003) by Métis playwright Marie Clements, enacts the non-linearity and circularity of Indigenous (particularly Dene and Métis) concepts of time, allowing a reconceptualisation of futurity based on 'historical consciousness' (Donald 2012). *Salmon is Everything* (2014) by Theresa J. May (non-Indigenous), Gordon Bettles (Klamath), Suzanne Burcell (Karuk), Kathleen McCovey (Karuk), Jean O'Hara (non-Indigenous) and the Klamath Theatre Project told stories of human/more-than-human relationality in Indigenous worldviews after a large-scale salmon kill in the Lower Klamath River (USA) in 2002. Through an Indigenous/non-Indigenous collaborative, theatre-making process, the performance focused on Klamath River people's understanding of the land and the more-than-human as relational rather than resource, developing reciprocity and responsibility. *Sila* (2014) by Chantal Bilodeau (non-Indigenous) includes multiple and intersecting stories of the Arctic, focusing on climate justice and Inuit culture and ecologies.

Resisting a generic and singular Indigenous dramaturgy, these works offer a multivocal and nuanced view of Indigenous ecologies as pluralities that provide necessary decolonising perspectives on the Anthropocene.

As I previously detailed at the beginning of this book, I was born on traditional Anishinabewaki territory in what is now called Cambridge, Ontario, Canada. I am of white European settler/immigrant ancestry (Belgian, Polish, Ukrainian, British). I am now an immigrant myself, living in the UK. This means I was born into a legacy of settler colonial violence as a white Canadian and I am also at risk of becoming complicit in a structure with a history of exploitation in the name of 'research' as an academic at a UK university. I acknowledge these complicated contexts and histories, rather than shy away from them, as I strive to approach ecodramaturgies from an intersectional, feminist, anti-racist, decolonised perspective. My approach is inspired by Palyku law academic Ambelin Kwaymullina's theory of listening (2018). She writes that white settler ecofeminists need to listen to Indigenous women if the movement will ever be inclusive of Indigenous Peoples. 'My suggestion is that non-Indigenous scholars must respect Indigenous sovereignty and meaningfully enact this respect, including through the layered process of listening to the voices of Indigenous women' (Kwaymullina 2018:193). I follow Kwaymullina's call to listen by centring Indigenous voices and by interrogating my own privilege, coming from white settler ancestry and complicit in the ongoing effects of colonialism. These ways of listening mean I draw on Indigenous ecological thought through the words of Indigenous scholars, activists and playwrights as a form of 'citational relational' (Ahmed 2014).

My aim here is not to appropriate Indigenous ideas or speak for any group of people; rather, I want to think critically about some of the underlying ethnocentric assumptions made about ecology and the knowledge systems that produce them. German-American community performance artist and researcher Petra Kuppers has argued for reading strategies of transnational performance practices in Indigenous/settler collaborations, specifically citing 'relational living, in the flow of history, speaking from webs of more than one voice, and attending to gaps' (2014: 5). By positioning herself in a living relationship with the performance practices she encountered, Kuppers attempts to decentre colonialist-marked methods of interpretation, of truth-seeking and performance criticism. Similarly, my own way into this research, as a non-Indigenous person, is not through claims of authority, but rather through suggestions of the way relational, complex and heterogeneous ecological knowledges, practices and activism offer ways of critiquing and disrupting the power structures and organisation of knowledge that govern our understanding of ecological relationships in theatre and performance.

The inherent interconnections between colonialism and ecological subjugation are too prevalent not to be considered by anyone working in ecology and art, and as Maori scholar Linda Tuhiwai Smith writes, the responsibility to make change is 'in both the non-indigenous and indigenous worlds' (2012: xii). I am also inspired by Quechua scholar Sandy Grande's *Red Pedagogy* (2004), which seeks out new epistemologies and paradigms based on Native American knowledges in education, which can open up and diversify critical theory in the academy, not towards assimilation of knowledge but towards decolonisation, recognition of difference and multiple ways of knowing. It is from this position that I approach decolonising the Anthropocene and ecodramaturgies more broadly.

Indigenous Ecodramaturgies

'The Crying Indian' was a widely viewed Public Service Announcement (PSA) in the US in the 1970s as part of the anti-litter campaign Keep America Beautiful. In the PSA, a 'Native American'[4] character paddles a birch bark canoe down a river. He is dressed in a buckskin outfit with colourful beading and fringe, his black hair in two long plaits, with a large white feather tucked into one side. He paddles past large industrial shipping vessels and smog-emitting factories, which dwarf his small canoe. In a booming baritone voiceover, we hear that 'some people have a great love and respect for this country, while others do not'—a point that is emphasised with a rhythmic drum-based score. He then walks ashore to discover the roar of a busy motorway where a bag of fast food rubbish is thrown from the window of a passing car to land at his feet. The camera zooms in on his face while a single tear slowly rolls down his cheek. Although this PSA has not been distributed widely since the 1970s, it has played a role in conditioning an image relevant to the cultural representation of Indigenous ecologies today: the pervasive trope of the 'ecological Indian'.

The story of 'The Crying Indian' is a typical story of settler colonial paradigms of representation. Indigenous peoples are used in a tokenistic way to further a political agency they had no input in shaping. They are erased as the part of the assumed audience of the PSA, and instead patronised through inaccurate representations of their values and way of life.

They are not materially compensated for the portrayal as a non-Indigenous actor is cast in the role and paid for his work.

The 'ecological Indian' is a stereotype often employed to suggest a type of ecological nostalgia, reframing colonial notions of 'otherness'. Non-Indigenous scholars Birgit Däwes and Marc Maufort write of this common stereotype:

> the "ecological Indian", or the keeper of a planetary spirituality, have proven tenacious and difficult to overcome ... Indigenous people are presented as noble custodians of their environment; instinctively harmonious with their environment, yet unable to meet the challenges of technology and civilisation. (Däwes and Maufort 2014: 12).

This eco-Indian trope simplifies multiple and complex cosmologies and relationships into a reductive and convenient stereotype that allows for the dismissal and marginalisation of Indigenous peoples as naïve and stuck in the past, rather than technologically advanced, urban-dwelling and modern, without critiquing its underlying assumptions. Not only does the trope misrepresent Indigenous cultures, it also perpetuates the romantic myth of nature as a balanced and harmonious system, ignoring the complexities and context of our current ecological situation. Däwes and Maufort claim that 'the images of the eco-Indian or eco-Aboriginal—effectively revived towards the end of the twentieth century—have been powerful instruments of dispossession and displacement' (2014: 12). Through this 'primitive' coding, colonial power and knowledge structures are uncritically reaffirmed rather than challenged. To counter this, Indigenous perspectives and decolonised ecologies can problematise and critique the Western-centric concepts of ecology and human/more-than-human relationships, attending to the underlying ideologies and structural violence. This is the case in the plays and movements discussed below: Idle No More reaffirms Indigenous rights, sovereignty and presence; *Burning Vision* (2003) counters dominant Euro-Western concepts of temporality and relationality; through a collaborative process, *Salmon is Everything* (2014) stages human and more-than-human community; and *Sila* (2014) centres Indigenous climate justice. There is a danger here, as a non-Indigenous person, of perpetuating the eco-Indian trope. For performance scholar Jill Carter (Anishinaabe/Ashkenazi), this risk is about further extraction of Indigenous ideas, art, objects and bodies by settlers. Therefore 'dramaturgical strategies that combat extraction, obstruct penetration, and thwart satiation' are needed (Carter 2020: 17). I am mindful of this risk as I look to platform Indigenous scholars, activism and artists in my approach to Indigenous ecodramaturgies.

Whereas the eco-Indian trope is unspecific and codes Indigenous knowledges as illegitimate, Indigenous ecodramaturgies focus on processes of making, stories and representations of distinct Indigenous peoples, communities and lifeways. The works in this chapter manifest specific Indigenous ecological knowledges, which critique colonial concepts of ecology, nature, land and time. Dione Joseph writes that 'dramaturgy, therefore, within an Indigenous context, can also be seen as a performed discourse connecting people with their own and each other's lived lives directly and immediately while allowing the external expression of the human condition and its interrelations to be produced within a context highly specific to its own Indigenous ontology' (Joseph 2019: 137). Indigenous ecodramaturgies, in specific and different ways, offer a decolonised perspective on the Anthropocene by acknowledging relationality, the power of stories and challenging colonial constructions of time and identity.

The Anthropocene: Which Humans Do We Mean?

The term Anthropocene, the geological epoch of the human, was coined by atmospheric scientists Paul Crutzen and Eugene Stoermer in 2000. It has become a key term and buzzword in much ecological discourse. As a geological epoch it is still being debated. Geographers Lewis and Maslin argue that humans have an unprecedented influence on the earth creating 'planetary-scale changes [that] rival any in Earth's geological history' (Lewis and Maslin 2018: 215). The Anthropocene is 'at its core this combination of the Greek words for "humans" and "recent time", means that the scale of human affairs is increasingly dictating the future of the only place in the universe known to harbour life' (Lewis and Maslin 2018: 215). The scale of human influence is leaving geological evidence and traces on the earth. However, this is not the whole story. The ways in which humans are (historically and currently) differentiated are not accounted for in this narrative.

Lewis and Maslin enter into the debate of when the shift to the Anthropocene took place. In order for a new epoch to be determined, they are looking for a Golden Spike. They argue that there was such a spike: the Orbis Spike of 1610, which involved settler colonial violence and 'New World'—'Old World' biotic exchange that is significant enough to consider it as a beginning of the Anthropocene:

> The Orbis spike implies that colonialism, global trade and coal brought about the Anthropocene. Broadly, this highlights social concerns, particularly the unequal power relationships between different groups of people, economic growth, the impacts of globalized trade, and our current reliance on fossil fuels. The onward effects of the arrival of Europeans in the Americas also highlights a long-term and large-scale example of human actions unleashing processes that are difficult to predict or manage. (Lewis and Maslin 2015: 177).

By framing settler colonialism as an 'example of human actions' that impact the earth, they are erasing the difference between humans, although they do acknowledge the unequal power relations between different people. As British author Richard Kerridge suggests 'this all-inclusive personified "humanity" is a construct that hides political questions about specific responsibility for environmental disasters' (2016: xv). The genocide of settler colonialism is not merely one example of how humans influence the earth, it is a specific form of violent domination of peoples, lands and more-than-humans that has shaped discriminatory structures and governance enacted today.

Political ecology scholar Audra Mitchell sees potential for a colonial critique of the Anthropocene in Lewis and Maslin's Orbis hypothesis, suggesting that it extends beyond geophysical properties to human bodies:

> The wounds inflicted by colonization do not pass away with the human bodies it directly enslaves; they are engrained in, and integral to, very lively Earth systems that persist today and will continue far into the future. Moreover, the hypothesis suggests that this legacy of violence has produced global conditions which ultimately encompass and affect all life-forms on Earth. (Mitchell 2015: n.p.)

Expanding this idea, colonialism then is engrained, not only in cultural and governmental systems, but also in human bodies and memories, cultural and social histories and the earth itself. Colonialism shapes structures and ideologies of ecological relationships. As Adams and Mulligan note 'it also transformed nature, creating new landscapes, new ecologies and new relations between humans and non-human nature; in the process, it created new ideologies of those relationships' (Adams and Mulligan 2002:

1). Settler colonial violence, genocide and domination 'changed the very categories within which nature and society were conceived' (Adams and Mulligan 2002: 5). There was not one incident of colonial violence, for example 1610, as Kwaymullina points out 'the colonial apocalypse was not one but many—a cataclysm of violence that began anew wherever colonizers arrived in the homeland of an Indigenous nation' (Kwaymullina 2018: 195). Colonialism is not just a 'spike' to mark a change in the earth's strata, it set up a violent foundation for how 'we' in the Western world understand and structure socio-ecological relationships, ecological epistemologies and the more-than-human world. Human history is ecological or 'natural' history, and these histories are shaped by unequal power relations, violence and domination. Understanding this enmeshment reveals how the Anthropocene story replicates colonial logic.

Crutzen and Stoermer suggest the introduction of the steam engine in 1784 marked the transition to the Anthropocene (2000: 18). However, as ecocritical scholar Joni Adamson points out, 'steam technology was not adopted by some natural-born deputies of the human species but by a tiny minority in Britain ... At no time did the species as a whole vote for a fossil fuel economy or exercise any shared authority over the destiny of the Earth systems' (2016: 160). This mirrors ecologists Andreas Malm and Alf Hornborg's critique of the Anthropocene, in which they refer to the 'a clique of white British men literally pointed steam-power as a weapon—on sea and land, boats and rails—against the best part of humankind, from the Niger delta to the Yangzi delta, the Levant to Latin America' (2014: 64). The popularity of the discourse, Adamson contends, has created 'a universalism that obscures vital social fact' (2016: 161). By not acknowledging the power structures that have produced the conditions of the Anthropocene, it erases differences between people, nations and collectives, which it collapses and conflates all into a universalising homogeneous species.

Geographer Kathryn Yusoff rewrites the origin stories of the White Geology of the Anthropocene. In her work *A Billion Black Anthropocenes or None* (2019), she argues that the Anthropocene not only ignores the effects of settler colonialism but also its racialisation. The story of the Anthropocene, she asserts, is premised on settler colonial violence and slavery, making it anti-Black. These ideologies are in the materiality of geology. The future tense of the Anthropocene means it elides 'recognition of the extinctions already undergone by black and indigenous peoples. Following in the wake of humanism, the production of the Anthropocene is predicated on Whiteness as the color of universality'

(2019: 51). Yusoff is not offering a corrective alternative to the Anthropocene with a billion Black Anthropocenes; she is offering a re-description towards 'more accountable, decolonized, geosocial futures' (2019: 61). Race structures the Anthropocene; therefore, to address the ecological crisis we find ourselves in, we must overcome extractive colonialism and acknowledge the '*economy of flesh* that underpins geologic practices … a geophysics of flesh that is Black and Brown' (2019: 61–62). She argues that to acknowledge the racialisation of the Anthropocene means to reframe the 1610 Orbis hypothesis origin story; 1610 misses the change in ecologies of 1452 when African slaves were forced into labour on sugar plantations in Madeira. The shift to land as resource and humans as labour underpins the logic of the Anthropocene(s). Yusoff's formulation of race in the Anthropocene discourse is essential to decolonised ecologies as it furthers understanding of the exclusions and erasures within the grand narrative.

Indigenous ecodramaturgies can counter some of the underlying ideology of ecological relationships to rewrite, reframe or resist the White Geology of the Anthropocene. Métis artist and anthropologist Zoe Todd details the role Indigenous art plays in a resistant praxis: 'In order to resist the hegemonic tendencies of a universalizing paradigm like the Anthropocene, we need joyful and critical engagement through many forms of praxis. I see Indigenous thought and practice—including art—as critical sites of refraction of the current whiteness of Anthropocene discourses' (Todd 2015: 252). Following Todd, Indigenous ecodramaturgies is a praxis, focusing on theatre and performance processes and practices, to decolonise the Anthropocene.

Idle No More

Burning sage, colourful speaking drums, multiple generations, traditional and contemporary clothing, people holding hands in a circle, moving their feet to the drumbeats, waving placards and flags, while snow is falling. The images of performances from Idle No More featured the round dance, which was performed at shopping centres and intersections in Canada and the US. SkyBlue Mary Morin's poem, 'A Healing Time', highlights the power of dance in the Idle No More movement:

> We dance
> to soften the hard lumps
> that have formed
> in the heart,
> the hurt inside...
>
> Indian women
> this dance is theirs
> to show respect
> the men sing for them.
> Never lifting the feet
> off the trodden ground
> for they must stay
> close to Mother Earth
> for they are one
> with her.
> (Morin 2014: 7–9)

The grassroots, youth-driven protest movement countered colonial socio-ecological violence through collective performances of Indigenous peoples in North America.

The Idle No More (INM) movement began in 2012 in Canada. For Greta Gaard, it is an example of how 'indigenous feminists have long advocated for gender, species, and environmental justice' (Gaard 2016: 91). The movement began with the remit to protect land and water. Idle No More was formed in response to then Canadian Prime Minister Stephen Harper's budget bill (Bill C-45) which diminished environmental protections of rivers on First Nations, Inuit and Métis land and challenged their sovereign rights (by rolling back the Navigable Waters Protection Act of 1882). This made it easier for industrial building projects and pipelines from the Alberta tar sands, such as the Enbridge Northern Gateway Pipelines Project (Gaard 2016). INM is explicitly working against settler colonialism:

> The impetus for the recent Idle No More events, lies in a centuries old resistance as Indigenous nations and their lands suffered the impacts of exploration, invasion and colonization. Idle No More seeks to assert Indigenous inherent rights to sovereignty and reinstitute traditional laws and Nation to Nation Treaties by protecting the lands and waters from corporate destruction. (Idle No More n.d.)

The movement was founded by three Indigenous (and one non-Indigenous) women: Nina Wilson (Nakota and Plains Cree from Treaty 4 White Bear territory), Sheelah McLean (non-Indigenous), Sylvia McAdam (Saysewahum) and Jessica Gordon (from Pasqua Treaty 4 Territory). They used performance as well as teach-ins, rallies, protests and social media campaigns to build the global movement.

Dancing is foundational to the ongoing Idle No More movement as the round dance takes a central place in INM gatherings. Anishinaabe/Métis writer Ryan McMahon describes the round dance flash mobs as an inclusive protest form with revolutionary potential (Fig. 6.1). There was a round dance action in the West Edmonton Mall (the largest shopping centre in North America), during the Christmas shopping period in December 2012 and then one in the Mall of America, Minneapolis in January 2013. The round dance is led by drums as 'the drum beat represents the heartbeat of Mother Earth' (McMahon 2014: 100). Through the round dance flash mobs, 'our communities are slowly regaining their strength … *A Round Dance Revolution*. It has reinvigorated and re-inspired our People.

Fig. 6.1 Idle No More demonstration, Dundas Square, Toronto, 2013. (Photo: Victor Biro/Alamy Stock Photo)

It has lifted the spirits of thousands' (McMahon 2014: 100). Anishinaabe writer Niigaawewidam James Sinclair's poem, 'Dancing in a Mall', captures the round dance under fluorescent light:

> writing with our feet
> we speak.
> in the air
> conditioning...
> so in the winter we
> danced
> we spoke we walked we danced
> we dreamed
> and we said no more
> (Sinclair 2014: 149)

By bringing the round dances in commercial, capitalist spaces there is a gentle critique of business-as-usual for most North Americans. Naomi Klein writes that with 'eagle feathers upstaging the fake Santas' (Klein 2014: 221), the incendiary global capitalist system, underpinned by colonialism, is momentarily unsettled. The critique is gentle because it is a performance that invites people to join in, take hands, follow the strong drum beat and build co-resistance by being together in an anti-capitalist way.

The round dances of Idle No More are part of what Todd calls the promise of Indigenous scholars and artists 'to speak back, reshape, and change the direction of current human-centric and Eurocentric framings of the Anthropocene' (Todd 2015: 249). This is 'an effective art of the Anthropocene' as it 'directly engages with the structural violences of heteropatriarchy and white supremacy as they shape discourse and praxis' (Todd 2015: 249). The women-led movement centred Indigenous sovereignty, without making 'Indigenous' peoples a monolith, in an intersectional way. The trauma of environmental collapse is not new for many Indigenous peoples; they have been living it since the beginning of settler colonialism. Chickasaw poet and novelist Linda Hogan writes that Indigenous nations have assumed extinction stories which creates a disconnection to the more-than-human world:

> We have taken on the story of endings, assumed the story of extinction, and have believed that it is the certain outcome of our presence here. From this position, fear, bereavement, and denial keep us in the state of estrangement from our natural connection with land. We need new stories, new terms and

> conditions that are relevant to the love of land, a new narrative that would imagine another way, to learn the infinite mystery and movement at work in the world. (Hogan 1995: 94)

The INM movement helped to create new stories for environmental justice that is based on an Indigenised way of knowing and living in the more-than-human world; new stories of life, culture and collective action. For McLean, one of the founders, 'the Idle No More movement is re-storying Canada—using public gatherings and mass media, it is actively re-telling stories which have been silenced, minimized, and denied, but also provides multiple forums to share stories that inspire hope and promote social and political change' (McLean 2014: 93). Through the mass protest movement, INM changed the narrative of Indigenous oppression and recuperated stories, finding new ways of being and resisting together for Indigenous peoples across Nations as well as settler allies.

Choctaw scholar LeAnne Howe writes of Native stories as living theatre, which makes connections and interrelations between peoples, time and places, as stories are performances of histories, beliefs and epistemologies. 'Native stories are power. They create people. They author tribes' (Howe 1999: 118). Stories create, shape and reinforce ideas, cosmologies and relations to the world, and have effects on understanding the past, present and future. Stories, then, are enacted as dramatic performances. Algonquin/Irish theatre-maker Yvette Nolan writes of Indigenous theatre as acts of medicine because they can negotiate solidarities and reconnect and build communities. 'The act of staging these things reconnects who we are as Indigenous people with where we have come from, with our stories, with our ancestors' (Nolan 2015: 3). Ecodramaturgies can engage an ecological imagination, grounded in a nuanced understanding of Indigenous ecologies that is relational, complex and pluralistic.

Burning Vision

Burning Vision (2003), by Métis playwright Marie Clements, was first performed at the Firehall Arts Centre in Vancouver in 2002, directed by Peter Hinton (non-Indigenous), followed by a national tour of Canada (and many awards) in 2003. The poetic play includes interconnecting stories, based on an Indigenous worldview that manifests temporal and spatial non-linearity. The play tells the story of the making of the first atomic bomb, tracing the history through uranium ore. White settlers 'discover' uranium under Great Bear Lake and open the world's first uranium mine on Sahtu Dene First Nations land in the Northwest Territories of Canada. They employed Dene men as ore-carriers who often died of cancer. The ore was then transported down the Mackenzie River and sent to test sites for the Manhattan Project in New Mexico. It was then detonated over Hiroshima in 1945 as the first atomic bomb.

The play is in four movements, tracing the uranium ore across four nations (Sahtu Dene Nation, Canada, US and Japan) and through four elements: from the earth, through water, over land and then into fire as an atomic bomb. The history is not traced in a linear order; rather it connects places, people and historical moments across time and space, through a complex web of materials and relations decades before and after the atomic bomb. Theresa May writes that 'the indigenous viewpoint from which the play is written, and which it enacts, allows for simultaneity of past, present and future, in which the spirit world co-exists with the embodied world, in which nothing is inanimate' (May 2010: 7). A Japanese fisherman feels the impact of unearthing uranium in the Northwest Territories in a different time. As non-Indigenous scholar Helen Gilbert points out:

> Indigenous communities in various regions have been disproportionately affected by nuclear expansionism ... indigenous peoples have been prominent among those suffering from illnesses caused by uranium extraction, trans-shipment and waste disposal, while their land has borne the brunt of environmental degradation. (2013: 197)

In *Burning Vision*, the connections between peoples subjected to marginalisation are enacted through materials such as flour, ore, bread and water, which are animated over timescales, producing effects that resonate over long distances. In Movement One, Rose, a young Métis woman who bakes bread, carries a large bag of flour, which slowly leaks a white circle of powder. In Movement Two, the flour circle becomes the compass and we see the Radium Painter, who paints the compass face with radium, licking the tip of her paintbrush and slowly poisoning herself. Then the white powder becomes radium dust, contaminating the ecosystem as children play with it and caribou eat plants covered with it. May refers to this as 'time-outside-of-time' in which 'the play makes visible a web of human agency that binds together places, people, and creatures' (2010: 6). Acknowledging this simultaneity opens up ways of critiquing ecological relationships and Eurocentric understandings, in which the animated materials and impacts of the past are felt as keenly as those in the present and future.

The simultaneity or circularity extends to human identities, as Däwes writes, 'within these permeable spaces and temporal environments, human identities—both Native and non-Native—are constructed as fluid and flexible, far from any essentialism, and with the full responsibility of acknowledging their ecological embeddings' (2014: 41). While *Burning Vision* includes multiple characters from different times and cultures, it also enacts Métis and Dene worldviews in which communication happens over multiple timescales with characters sharing time and space across 'historical periods' and geographical distances. The character of the Dene Seer (a medicine man) elaborates this: 'maybe we are all talking at the same time because we are answering each other over time and space. Like a wave that washes over everything and doesn't care how long it takes to get there because it always ends up on the same shore' (Clements 2003: 65). This circularity can also be a resistance to the dominant colonialist narrative of history, as it enacts events through Indigenous viewpoints revealing the interconnections of people, land and more-than-humans across time and space. Although the play text includes a linear timeline of events of uranium ore from the late 1880s prophecy of the Dene Seer to 2002 when the play itself premiered, this is merely a helpful reference to the reader, as the play does not follow this linear construction of history.

Unangax scholar Eve Tuck and non-Indigenous scholar K. Wayne Yang argue that 'the too-easy adoption of decolonizing discourse (making decolonization a metaphor)' enacts a series of tropes 'which problematically attempt to reconcile settler guilt and complicity, and rescue settler futurity' (2012: 3). Rather, Indigenous futurity needs to be the focus of any decolonisation practices. I argue this concept of futurity is not solely forward looking, but is based on a non-Eurocentric concept of the future that is connected to the past and present. *Burning Vision* enacts a historical consciousness, where, as Todd writes, 'the past occurs simultaneously in the present and influences how we conceptualize the future. It requires that we see ourselves related to, and implicated in, the lives of those yet to come' (Todd 2015: 250). These are what Dwayne Donald calls 'survival stories that give life back to those of us living today' (Donald 2012: 545) by centring Indigenous futurity, as many Indigenous peoples have faced the possibility of extinction. The characters in *Burning Vision* embody the material histories and bring them into the present and future. The character of Little Boy is described as: 'A beautiful Native boy. Eight to ten years old. The personification of the darkest uranium found at the centre of the earth' (Little Boy is also the name of the atomic bomb dropped on Hiroshima), while Rose is 'A Métis woman in her twenties who walks between Native and non-Native lines as she works in the North at her father's Hudson's Bay Store. A bread-maker and a dreamer looking for her place' (Clements 2003: 1–2). The characters embody the physical effects of mining and painting with radium as well as the material history, present and future of settler colonialism and oppression (including the Japanese internment in the US and Canada in WWII) and terror that led to the many deaths of the building, testing and ultimately firing of the atomic bomb. The poetic rather than naturalistic style of the piece allows the historical consciousness to anchor the play, allowing us to see ourselves as 'implicated in the lives of those yet to come', both humans and more-than-humans.

Kwaymullina writes of the 'narrative sovereignty' of Indigenous peoples, or a freedom to tell stories outside of colonialist marked frames, 'whereby belonging is grounded in story and particularly in the stories of the Ancestor beings' (Kwaymullina 2018: 198). In *Burning Vision*, the stories of belonging are connected to ancestral relations and beings. Characters are doubled in different places and times; elements carry ancestral history across geographical borders. In Movement Four, Little Boy embodies the Dene Seer (who is heard through voiceover) as he describes his vision, an eerily accurate prophecy of uranium mining on Dene land, resulting in the hurting and burning of people who look like the Dene. Within the four movements of the play, characters across different nations and times are interwoven. Bookended by the atomic bomb blast—which 'lasts a long time and reaches far into the distance, until at last the explosion is complete and it is quiet. Darkness' (Clements 2003: 8)—the characters of the play seem to be called forth into the action by The Widow, an older Dene woman who is keeping a fire in memory of her late husband, a Dene ore-carrier. This blurring of space and time allows for de-centring of whiteness as alliances are drawn between the cancer-ridden Dene ore-carriers and the Japanese victims of the atomic bombs, rather than the white settlers. The Dene seer says, 'The people they dropped this burning on ... looked like us, like Dene' (Clements 2003: 111). Both the Dene and the Japanese embody the scars and effects of the ecological exploitation and colonialist violence.[5]

In her study of Native American drama, non-Indigenous scholar Christy Stanlake identifies Native dramaturgy as frequently non-linear: 'like storytelling, Native plays often utilize a non-linear, sometimes cyclical, plot structure' (2009: 23). This non-linearity can also be a resistance to the dominant colonialist narrative of history, as it enacts events through Indigenous viewpoints, revealing the interconnections of people, land and more-than-humans across time and space. As Chief Looking Horse articulates, 'in a Sacred Hoop of Life, there is no ending and no beginning' (Looking Horse 2018). This concept of non-linearity also foregrounds the cultural construction of timekeeping and organisation, the prioritisation of 'efficiency', which is laden with value judgments that privilege one culture over others that may organise time differently (i.e. Indian time). In *Burning Vision*, the dramaturgy is non-linear with characters sharing the stage in different times and places. Western perceptions of time are troubled through ecological conditions which call forth the past, connecting it materially with the present and future. The Widow says in the fire: 'We used to be able to tell where we were by the seasons, the way the sun placed itself or didn't, the migration patterns of the caribou. Time' (Clements 2003: 33). A caribou herd heralds the transitions between movements, marking a different kind of time, and at the end of the play after the atomic bomb blast, 'glowing herds of caribou move in unison over the vast empty landscape as cherry blossoms fall until they fill the stage' (Clements 2003: 114). A circular and relational understanding of ecology may be able to resist or delimit neoliberal modes of timekeeping and acknowledge the material effects of past and current actions on the future. In this way, non-linearity and the inseparability of past, present and future in Indigenous ecological epistemologies is a potent framing of ecodramaturgies towards decolonisation.

Salmon is Everything

A group of faculty, students, staff and community members from Humboldt State University, followed by University of Oregon, undertook a Native/non-Native collaboration for the community-based performance *Salmon is Everything* (2014). Indigenous ecodramaturgies in this piece took the form of collaborative process and community building, while centring the Indigenous peoples of Klamath River and their worldviews. In 2002, there was an unprecedented fish kill, a large-scale die-off that decimated the chinook and coho salmon population, on the Lower Klamath River (running through California and Oregon), which was harmful to local people who depended on the river for their livelihood: Indigenous Nations who fished salmon (primarily Karuk, Yurok, Hupa,

Wiyot, Modoc and Klamath Tribes), commercial fisherman, and farmers and ranchers who depended on the water (May 2014a: 201). Julie, a Yurok-Karuk character in the play, underscores the fish kill as a wider issue: 'This is a community issue, not just an Indian issue. Most non-Natives see it as an Indian issue, and they don't even understand what that means. It's a spiritual issue for all of us' (May and the Klamath Theatre Project 2014: 65). Gordon Bettles, member of the Klamath Tribes and cultural advisor on the University of Oregon production of *Salmon is Everything*, writes that the historic fish kill is related to settler colonialism, 'caused by a collision of cultures that began when the first ships landed on the east coast of North America in 1492' (2014: xi). The fish kill was a result of 'high water temperatures, low water levels, and toxic algae levels caused by the overuse of water by agriculture' (Bettles 2014: xi). Native scientists warned the Environmental Protection Agency and other agencies of these material factors affecting salmon in 2001, but pressure from farmers and ranchers caused them to reverse protections on the salmon. Again, settler colonial values are protected and reaffirmed through governance while the cultural and spiritual significance of the salmon to the Indigenous peoples of the area is effectively overruled and seen as less important.

Director Theresa May wanted to initiate a performance project involving the communities impacted by the fish kill because she thought it would be a way to bring out the Native voices that were seemingly absent from the public discussion of the crisis. This collaborative theatre-making process echoes what Kuppers describes as a way of 'speaking from webs of more than one voice, and attending to gaps' (Kuppers 2014: 5). May writes that, at the first meeting to discuss the project at Humboldt State University in California, a Native Elder in attendance stopped her description of the project, saying 'What's all this talk about telling an Indian story when we're sitting in a classroom of a racist institution?' (2014b: 104). In that moment, the Elder cut to the heart of the anomaly of Indigenous/non-Indigenous collaborations. How can institutions and knowledge frameworks laden with Western value judgements and modes of knowledge legitimation based on colonialism engage with Indigenous ecological epistemologies in a way that does not reproduce colonial subjugation and marginalisation? Any decolonising praxis needs to acknowledge the legacies, systems and structures of colonialism and neocolonialism. I have entered into this research mindful of the critique of generalising and reinforcing Western modes of knowledge production, while attempting to think in dialogue with Indigenous ecological knowledges and how they contribute to, critique and fortify ecological theatre and performance.

For May, the exchange with the Elder began a process of listening to Native stories, involving Karuk, Yurok, Hupa, Modoc and Klamath Tribes artists, students and community members in the process of making the performance as collaborators and deferring to Karuk, Klamath Tribes and Wasco Nation cultural advisers (Suzanne Burcell and Kathleen McCovey at Humboldt State and subsequently Gordon Bettles and James Florendo at the University of Oregon). They agreed that the university would have no ownership over the stories shared. This process helped to ensure the 'end product reflected the values of the Native community' (May 2014b: 104). This is what Nolan refers to as an 'eighth fire production', which is 'created by a group of Indigenous and settler artists in an attempt to create understanding and forge a new and healthy way forward together' (2015: 21). The process also included learning about Yurok, Karuk, Hupa, Wiyot, Modoc and Klamath Tribes cosmologies and spiritual life, the meaningfulness of traditional objects (not as props), and of negotiating the historical tensions between the Indigenous Nations on the watershed and the government institutions, large-scale agricultural farmers and university researchers. This also included learning about the watershed and river itself. The character Will (a Yurok-Karuk fisherman) describes the river in the play 'as part of me, the lifeblood of my people…The Klamath is my home, my church, garden, highway, counsellor, friend, brother—hell, provider' (May and Klamath Theatre Project 2014: 74). Through these tensions and learning process, the project brought performance-making strategies and techniques to bear on a controversial local ecological issue, centring the Klamath River people's worldview while staging affective stories of the river not usually represented in mainstream public forums.

However, one of the dangers of this kind of cultural representation by non-Indigenous theatre-makers and audiences is placing Indigenous people as 'spokesperson' or 'cultural informant'. Nolan speaks of trying to 'resist the pressure to serve as a cultural informant, a guide to the Indigenous world (view) for the non-Native traveller' (2014: 111). It is imperative for non-Indigenous scholars and practitioners to understand 'the great abyss across which we work, as Natives and non-Natives' (Nolan 2014: 111). Working in collaboration to understand and centre different worldviews or cosmologies that are not 'primitive' but grounded in rich tradition while continuing to evolve as relevant knowledge is one of the ways we may start to cross 'the great abyss' towards decolonisation.

Kathleen McCovey (Karuk) was a cultural advisor for *Salmon is Everything* and wrote many of the lines for the character of Rose, a Karuk elder, despite having never worked on a play before. She writes in a commentary accompanying the script about her experiences working on the play and feeling that theatre was a powerful medium in which to communicate: 'We have to share the knowledge we have. For example, as a Karuk person who has grown up on the River, I have place-based knowledge about the land that I live on and the species with which I coexist' (McCovey 2014: 99). She felt empowered to share the stories and culture of her people, not as a cultural informant but as a co-creator of *Salmon is Everything*.

Salmon is Everything (2014) staged the cosmology of Klamath River peoples, which links identity to place and the more-than-human to form a sense of what it means to relate to the ecological world. Nolan, drawing on playwright Daniel David Moses, writes of protagonists in Native theatre as often being a community rather than an individual (2015: 41), which is different to dominant Eurocentric modes of storytelling in theatre. *Salmon is Everything* features a human and more-than-human community (the salmon, the river and all the life within it) as the protagonists and was authored by staff, students, and Klamath River community members, under the name The Klamath Theatre Project. The resulting play centres Klamath River people's stories and worldviews: the character of Julie (Yurok-Karuk) a teacher, and her extended family (partner, aunt, niece, nephew and cousin) who are Yurok-Karuk and Yurok, ground the story, which also features a family of Upper Klamath ranchers, fish biologists, farmers, tourists, a politician, reporter, priest and logger's daughter. The stories are performed through monologues and short scenes, first separating the different people and perspectives before they come together as an ensemble for a Town Hall meeting midway through the play and then again at the end for a ceremonial scene entitled 'Sacred Is'. The salmon are also central to the story, connecting the characters. The collage and overlap of stories, dialogue and people builds community between the different characters and perspectives in the play while also maintaining the Klamath River people's values for which the salmon and the river are sacred.

Through foregrounding community in the dramaturgy, an intricate story is formed of the multifarious effects of the fish kill on different characters. In the play, the Klamath River people's relationship to the-more-than-human is illustrated by Julie, who speaks of her Yurok-Karuk people: 'they are the trees, the water, the fish. That the Salmon are brothers is not some kind of myth; the Salmon are not symbols of life, they are life.' (May and Klamath Theatre Project 2014: 44). To understand the salmon as brothers positions the more-than-human not as a resource but as fellow living beings. The Yurok-Karuk people fish the salmon but recognise the need to do so in a way that respects the life of the salmon and sustains the salmon population for future years. This continuum of life counters the binary-making practices that separate human/nature and human/nonhuman, moving towards more ecological modes of being. This continuum of life can redress the boundaries and categorisations that privilege the Western conception of human and devalue the more-than-human, leading to ecological subjugation.

Todd describes the kind of art she wants to see following Indigenous principles of care and kinship with the more-than-human: 'I want art that enters my veins and comes pouring out like fish, stories about the river, struggling against the current…I want art that remembers that the stories we tell through it tie us to land and fish and dreams and past and present and future all at once and I want art that is attentive and tender to the stories that are told even in forms illegible to funding agencies and academic analyses.' (2017:105). Indigenous/non-Indigenous collaborations, like *Salmon is Everything*, can reframe colonial attitudes to more-than-human relations, telling stories that connect people to place and its many inhabitants. For the Klamath River peoples, salmon is kin and there is an ethical responsibility towards reciprocity with them. Telling these stories through Indigenous paradigms can act as a corrective to the erasures of the Anthropocene. The ecodramaturgical processes and stories centre Indigenous epistemologies that offer ways of resisting Eurocentric, colonial assumptions and practices tied to ecology, extractivism and resourcism.

Sila

Sila (2015) by non-Indigenous French-Canadian playwright Chantal Bilodeau (which premiered at Underground Railway Theater in Boston in 2014) focuses on climate change effects on the Arctic. Through interconnecting stories, we hear about the opening up of the Northwest Passage as an international trade route (now that the pack ice has melted) and the effects of climate changes on the Inuit and the more-than-human world. The play is named after the Inuit idea of *sila*, the spiritual relationship between people, climate and ecological processes, and considers it in relation to the politics of climate change and the Arctic. Within Inuit cosmology the concept of *sila* could be translated in a number of different ways, including as Timothy Leduc notes, 'a spiritual reference that is meant to contextualize the physicality of human relations within broader ecological processes like the weather' (Leduc 2010: 27). I saw a reading of this play at the Staging Sustainability conference at York University in Toronto in 2011 and again at the Earth Matters on Stage symposium at Carnegie Mellon University in 2012 (for which it won the Ecodrama playwriting competition). The play is set on Banff Island in the territory of Nunavut, Canada, and features seven interconnecting storylines about the current ecological and political context of the Arctic, including an Inuit climate change activist and her daughter and grandson, Canadian Coast Guard officers and a visiting climate scientist. The play centres Inuit characters and Inuit mythology to draw out the ways in which climate justice is needed for those oppressed and marginalised by colonial-marked ideologies of ecology.

The activist character, Leanna, is modelled on real-life Inuk climate change advocate and Nobel Peace prize nominee Sheila Watt-Cloutier. Within the play, we see how the Inuit are trying to raise global alarms about the melting glaciers and the effect of climate change on their culture and livelihoods. The Inuit are disproportionately affected by climate change despite their relatively small contribution to it. Not only are Inuit habitats being depleted due to climate change, Inuit health is impacted by pollutants far away from where they live. This collapsing of geographical distance is evident in the dioxins accumulated in their food sources, air and water (including bioaccumulating in breast milk as discussed in Chap. 3). As scholar Anne Lucas argues, 'researchers determined the majority of the pollution in the Canadian Arctic originated at solid waste incinerators, copper smelters, and cement kilns located in the midwestern United States' (Lucas 2004: 191). This mirrors the ecological effects of resource inequality on postcolonial nations, which generally have low carbon emissions (compared to the Global North) though they are disproportionately susceptible and endangered by the effects of climate change (DeLoughrey and Handley 2011: 26). It is also what Astrida Neimanis asserts as 'important mutations in forms of colonial power, where colonizers need not physically occupy a place with their discrete bodies for the environmental effects of (neo)colonial power to be felt' (Neimanis 2017: 164). In this way, a few, specific humans, operating within violent power structures, may be exerting ecological influence that harms not only the more-than-human world but marginalised people in other places. According to Lucas, even the reduction of emissions cannot alleviate the health impacts on the Inuit because 'the slow breakdown of dioxins in fatty tissues of humans and animals means emissions from the past and those that continue today will remain a significant health factor for years to come' (Lucas 2004: 192). This is an example of the way in which the past is marked on the earth and in the bodies of those marginalised. Watt-Cloutier is an international voice of the Inuit and argues that 'the Arctic is seen as a global barometer for climate change, and Inuit are responsible sentinels that have reached out to warn the world about this important issue that interconnects all of humanity' (Cape Farewell 2013: 33). Bilodeau writes her version of the Inuit activist as an international figure trying to draw the attention of the world to the Arctic, while faced with government inaction. The behaviour and lifestyles of the Global North and the colonialist relationship to the more-than-human world ostensibly have profound socio-ecological impacts on the Arctic and the Inuit who live there, yet it remains a 'foreign' and distant reality while the warnings and ecological knowledge of the Inuit are ignored or overlooked by policymakers.

Sheila Watt-Cloutier describes the Inuit's changing relationship to the land: 'the land that is such an important part of our spirit, our culture, and our physical and economic well-being is becoming an often unpredictable and precarious place for us' (2015: ix). This spiritual and cultural relationship is captured in the play and detailed in the playwright's notes at the beginning of the text, where Bilodeau writes of two Arctics present in the story: 'the Arctic of the Inuit and the Arctic of the Southerners. The Arctic of the Inuit is warm, raw, and fiercely alive. It feels and sounds like the music of Tanya Tagaq. The Arctic of the Southerners is cold, mystical in its foreignness, and rarefied' (2015: 8). The play is able to centre Indigenous voices, by including Inuktitut language spoken by the Inuk characters and a glossary of Inuktitut terms and phrases. It also includes Leanna's daughter Veronica, performing spoken-word poems by Inuk artist Taqralik Partridge (*Eskimo Chick* and *No Sleep for the Wicked*), and Inuit mythology. In these ways the eco-Indian trope is not substituted for Indigenous knowledge, epistemologies and voices.

In *Sila*, Leanna has filed a petition against the US government saying the human rights of the Inuit are being violated by conditions of climate change and the lack of action on it. The petition, which contained 'A hundred and seventy-five pages of thoroughly researched scientific facts and first-hand witness testimonies' (Bilodeau 2015: 34), and subsequent appeal is denied. This mirrors the fight for environmental justice for many Indigenous nations. For author Dina Gilio-Whitaker (Colville Confederated Tribes), federal laws and power structures that ignore

Indigenous rights and sovereignty is not a form of environmental racism but domination, as they do not take into account Indigenous relationships to the land based on relationality, reciprocity and responsibility (2019). She argues that non-Native people need to reconceptualise our relationship to the land based on these principles if we are going to survive the so-called Anthropocene. Environmental justice needs to be Indigenised, and the ecocide against Native peoples stopped. This is the continuing violence of settler colonialism. Lucas points out that the 'the failure of international and national bodies in examining risk for environmental harm of groups like the Inuit is firmly rooted in oppressive structures that categorise women, indigenous peoples, people of colour, and poor people as the "other," who clearly are not the "average" person regulations are enacted to protect' (Lucas 2004: 200). Laws are designed with this 'average' person in mind, who is usually coded as a white, Western man with the capitalist development imperative elevating profits over environmental harm and human rights (199), which Leanna (and Sheila Watt-Cloutier) is fighting against. The Inuit have the right to cultural self-determination, free from the colonial and capitalist narratives of development, which are presented in the play through the character of Thomas, a white Canadian Coast Guard officer who wants to build a port in the deep sea of the Arctic. He is in conflict with Leanna over it as she does not see how the port will be good for the Inuit—it is a refuelling station for military and civilian freighter ships only possible due to Arctic ice melting as a result of climate change. For Indigenised justice to occur for the Inuit, transnational corporations, who are largely responsible for the effects of climate change and toxins in the Arctic, and are enabled by nation states, need to be held accountable.

Anthropologist Julie Cruikshank observes that reductive assumptions of traditional ecological knowledge can be read as a recasting of colonial ideas of 'primitive superstition, savage nobility, or ancestral wisdom' that re-inscribe colonial domination by seeing Indigenous knowledge as 'an object for science rather than as a kind of knowledge that could inform science' (Cruikshank 2004: 21). The Canadian government seems to be ignoring the warnings of Indigenous nations about the effects of climate change in the Arctic. This could be partly because of colonial assumptions of the 'noble savage' associated with the eco-Indian trope that devalues and marginalises Indigenous ecological knowledges as primitive and irrelevant to contemporary society. The climate scientist in *Sila*, a French-Canadian called Jean, needs an Inuk elder to take him on his research expedition to fulfil the 'community' requirement of any research that takes place in Nunavut. Tulugaq, the Inuit elder, explains to him the role of traditional knowledge:

> That is *Inuit qaujimajatuqangit*. Inuit traditional knowledge. Old learning about living in peace with people, animals, nature. Arctic is not just numbers. Arctic is stories. Like *aqsarniit* story. *Qallunaat* learning: lots of numbers. But it comes here—(pointing to his head). Only here. Not good for us. Inuit *qaujimajatuqangit* comes here—(pointing to his head), here—(pointing to his heart), and here—(moving hands and feet). Inuit *qaujimajatuqangit* is alive. Observation, experience. Always changing. Numbers are not enough. We need stories. You understand? (Bilodeau 2015: 59–60)

Rather than being the object of study, *Inuit qaujimajatuqangit* [traditional knowledge] is considered a relevant (and lifesaving) study in and of itself. Tulugaq knows much more about the land and more-than-human world of the Arctic than Jean does, despite the fact that Jean is a famous scientist who has been studying the Arctic for 15 years. Tulugaq has to save his life after he falls through the ice, despite his warnings.

Inuit traditional knowledge is not patronised or dismissed as 'primitive'. In this way, the play avoids the trope of the eco-Indian with Tulugaq and Leanna. It depicts the struggle and conflict Leanna faces in being the international voice of the Inuit people. This requires her to travel a lot and spend little time at home, straining her relationship with her daughter Veronica. After missing an event at the school where Veronica teaches, an angry Veronica yells at her: 'I promised the students an exciting day with one of the most important leaders in their community and you didn't even show up! There's no excuse for that' (Bilodeau 2015: 63). She promises to be more present after the human rights violation appeal, but she believes what she is doing is necessary for her family, her community, her land. She is a complex and conflicted contemporary character, who is not without fault. She is not a one-dimensional stereotype or romanticised.

By the end of the play, despite dealing with death, loss and hardships, the Inuit community members realise their fight for climate justice is far from over. This commitment can be read as an understanding of climate justice as part of what it means to live and relate to the world in the context of global climate change and Indigenous marginalisation, or what Anishinaabe writer Gerald Vizenor refers to as 'survivance' (1999). This is Indigenous world-making that is 'more than survival, more than endurance or mere response; an active repudiation of dominance, tragedy, and victimry' (Vizenor 1999: 15). More than survival and resistance, this is a way of relating to the settler colonial power structures which inflict genocide and ecocide. Vizenor argues that Native stories are full of practices of survivance (2009: 1). The Inuit stories in the play relate survivance to climate justice, the way of relating for the Inuit is not just about survival but being a presence where there was absence and erasure, pushing back against the narrative of their extinction.

Sila presents Inuit mythology through a 'Western' theatrical form. The piece includes characters who enact Inuit mythology, including the story of Nanurjuk, a polar bear who went into the sky to become a star in the Orion constellation, and Nuliajuk, Inuk Goddess of the Ocean and Underworld. The Mother and cub polar bears have their own narrative and voice, rather than just standing in as a representation of climate change (or nature). In the 2012 and 2014 productions they were played by puppets, with the Mama polar bear voiced by the same actor who plays Veronica. Rather than climate change mascots (Doyle 2011), the polar bears had agency, although they were anthropomorphised as they spoke English and expressed human emotions. The polar bear cub dies, being unable to swim the distances required to find food in the new realities of melted ice due to climate change. The storyline parallels the loss of Veronica's son, who dies by suicide. The high rates of suicide in Indigenous youth in North America are part of the structural violence and domination of colonialism. The parallel storylines, linking the polar bears' loss of life and the Inuit youth's loss of life, connects the social and ecological effects of colonial ideologies of resourcism, extraction and capitalism, shaping ecological relationships. Although the play follows a linear time frame, and other Eurocentric modes of storytelling and theatrical conventions, it centres Indigenous voices and Inuit mythology. In the play, Inuit goddess of the ocean and underworld Nuliajuk visits Jean when he falls through the ice and gives the other human characters the strength to carry on with their lives and the campaign after the death of Veronica's son. In some ways, this is a convenient *deus ex machina* that solves a key problem and brings the story to an ultimately uplifting ending, which could be read as instrumentalising Inuit cosmology to service a Euro-centric storytelling form. However, this also acknowledges the role of Indigenous spirituality in practices of survivance.

By staging the complexities of the Inuit in the climate-changed Arctic, *Sila* engages a nuanced understanding of traditional ecological knowledges and cultures. For DeLoughrey and Handley (2011) literature, through imaginative means, can foreground the political and ethical force of an ecological worldview. Ecodramaturgies has this same opportunity to foreground Indigenous modes of ecological thinking with political and ethical dimensions to counter dominant narratives that reproduce colonial violence.

Ethical Imperatives for Decolonising Performance and Ecology

Stories in performance and the performance of stories help our understanding of the material effects of climate change across multiple timescales, refracted through multiple worldviews. In considering what a decolonised future might be in a climate-changed world, I have written elsewhere that 'theatre and performance can be a place where the relationship between nature and culture is negotiated, where futures are imagined and traditional knowledge is recuperated' (Woynarski et al. 2020: 204). Acknowledging difference while understanding commonalities, particularly in the global effects of climate change, is at the heart of Indigenous ecodramaturgies. The ethical demand and imperative exists for broad-reaching acknowledgements that reflect the diversity of knowledge and interconnections between settler colonialism, ecological subjugation and Indigenous oppression, while not appropriating marginalised voices. To seriously consider and reciprocally enact these connections, and their material impacts, requires engagement from Indigenous and non-Indigenous scholars alike: avoiding patronising codification or romanticised tropes. Although my aim here to is resist colonial frames of ecological knowledge, there is a danger of doing what Tuck and Yang call settler moves to innocence:

> those strategies or positionings that attempt to relieve the settler of feelings of guilt or responsibility without giving up land or power or privilege, without having to change much at all. In fact, settler scholars may gain professional kudos or a boost in their reputations for being so sensitive or self-aware. Yet, settler moves to innocence are hollow, they only serve the settler. (Tuck and Yang 2012: 10)

It is useful to stay with this discomfort and recognise that my decolonising aims are not easily reconciled by a non-Indigenous white settler. This is a process in continuous flux as settler and Indigenous thinkers continue to seek out non-hierarchical ways of engaging heterogeneous and historically marginalised forms of knowledge within the academy. Indigenous ecodramaturgies as multiple and evolving provide a critique of Eurocentric hegemonic ecological discourses of the Anthropocene and open up performance and ecology in a more inclusive and diverse way.

Notes

1. The Dakota Access pipeline is operated in conjunction with the Energy Transfer Crude oil pipeline. This system of pipelines leaked at least eight times in the first six months of operation in 2017. Water contamination was not reported, although soil contamination in the spill areas was (Brown 2018).
2. Some of the material in this chapter was published in different forms in 'Ecological Sentinels: Indigenous Heroes or Colonial Cliché?' in *Research in Drama Education: The Journal of Applied Theatre and Performance* (Woynarski 2015a) and 'Locating an Indigenous Ethos in Ecological Performance' in *Performing Ethos* (Woynarski 2015b).
3. When referring to specific Indigenous nations, and cultures, I follow the author's usage, including capitalising or not of 'Indigenous' in direct quotations.
4. The actor in 'Crying Indian' PSA was known as Iron Eyes Cody. He played a number of popular Native American roles in his career, although he was not of Native American descent but rather Italian-American.
5. This is not to imply that Japan has not been responsible for its own colonial violence.

References

Adams, W.M., and Martin Mulligan. 2002. *Decolonizing Nature: Strategies for Conservation in a Post-Colonial Era*. Abingdon-on-Thames: Routledge.

Adamson, Joni. 2016. We Have Never Been Anthropos: From Environmental Justice to Cosmopolitics. In *Environmental Humanities: Voices from the Anthropocene*, ed. Serpil Oppermann and Serenella Iovino, 155–173. London; New York: Rowman & Littlefield International.

Ahmed, Sara. 2014. White Men. *Feministkilljoys*, November 4. https://feministkilljoys.com/2014/11/04/white-men/.

Bettles, Gordon. 2014. Foreword: When Cultures Collide. In *Salmon Is Everything: Community-Based Theatre in the Klamath Watershed*, ed. Theresa J. May, xi–xvi. Corvallis: Oregon State University Press.

Bilodeau, Chantal. 2015. *Sila*. Vancouver: Talonbooks.

Bird, Michael Yellow. 1999. 'What We Want to Be Called: Indigenous Peoples' Perspectives on Racial and Ethnic Identity Labels. *American Indian Quarterly* 23 (2): 1.

Brown, Alleen. 2018. Five Spills, Six Months in Operation: Dakota Access Track Record Highlights Unavoidable Reality—Pipelines Leak. *The Intercept*, January 9. https://theintercept.com/2018/01/09/dakota-access-pipeline-leak-energy-transfer-partners/.

Carter, Jill. 2020. "My! What Big Teeth You Have!": On the Art of Being Seen and Not Eaten. *Canadian Theatre Review* 182 (Spring): 16–21.

Clements, Marie. 2003. *Burning Vision*. Vancouver: Talonbooks.

Cruikshank, Julie. 2004. Uses and Abuses of "Traditional Knowledge": Perspectives from the Yukon Territory. In *Cultivating Arctic Landscapes: Knowing and Managing Animals in the Circumpolar North*, ed. David G. Anderson and Mark Nuttall, 17–32. New York: Berghahn Books.

Crutzen, Paul J., and Eugene F. Stoermer. 2000. The "Anthropocene". *International Geosphere-Biosphere Programme Global Change Newsletter*, no. 41 (May): 17–18.

Däwes, Birgit. 2014. Stages of Resilience: Heteroholistic Environments in Plays by Marie Clements and Yvette Nolan. In *Enacting Nature: Ecocritical Perspectives on Indigenous Performance*, ed. Birgit Däwes and Marc Maufort, 21–46. Brussels: P.I.E. Peter Lang.

Däwes, Birgit, and Marc Maufort. 2014. *Enacting Nature: Ecocritical Perspectives on Indigenous Performance*. Brussels: P.I.E. Peter Lang.

DeLoughrey, Elizabeth M., and George B. Handley, eds. 2011. *Postcolonial Ecologies: Literatures of the Environment*. New York; Oxford: Oxford University Press.

Donald, Dwayne. 2012. Indigenous Métissage: A Decolonizing Research Sensibility. *International Journal of Qualitative Studies in Education* 25 (5): 533–555.

Doyle, Julie. 2011. *Mediating Climate Change*, Environmental Sociology. Burlington, Vermont: Ashgate.

Cape Farewell. 2013. *What Does Culture Have to Do with Climate Change? Everything*. Carbon 14: Climate is Culture Program Guide. Canada, December 29.

Gaard, Greta. 2016. Where Is Feminism in the Environmental Humanities. In *Environmental Humanities: Voices from the Anthropocene*, ed. Serpil Oppermann and Serenella Iovino, 81–98. London; New York: Rowman & Littlefield International.

Gilbert, Helen. 2013. Indigeneity, Time and the Cosmopolitics of Postcolonial Belonging in the Atomic Age. *Interventions* 15 (2): 195–210.

Gilio-Whitaker, Dina. 2019. *As Long as Grass Grows: The Indigenous Fight for Environmental Justice, from Colonization to Standing Rock.* Boston, MA: Beacon Press.

Grande, Sandy. 2004. *Red Pedagogy: Native American Social and Political Thought.* Lanham, MD: Rowman & Littlefield Internat.

Hogan, Linda. 1995. *Dwellings: A Spiritual History of the Living World.* New York: W W Norton & Co Inc.

Horton, Jessica L. 2017. Indigenous Artists against the Anthropocene. *Art Journal* 76 (2): 48–69.

Howe, LeAnne. 1999. Tribalography: The Power of Native Stories. *Journal of Dramatic Theory and Criticism* 16 (1): 117–126.

Idle No More. n.d. The Story. Accessed May 17, 2019. http://www.idlenomore.ca/story.

Joseph, Dione. 2019. Cradling Space: Towards An Indigenous Dramaturgy on Turtle Island. In *Performing Turtle Island: Indigenous Theatre on the World Stage*, ed. Jesse Rae Archibald-Barber, Kathleen Irwin, and Moira Jean Day, 131–145. Regina: University of Regina.

Kerridge, Richard. 2016. Foreword. In *Environmental Humanities: Voices from the Anthropocene*, ed. Serpil Oppermann and Serenella Iovino, xiii–xvii. London; New York: Rowman & Littlefield International.

Klein, Naomi. 2014. Awake, Hungry, and Idle No More. In *The Winter We Danced: Voices from the Past, the Future, and the Idle No More Movement*, ed. The Kino-Nda-Niimi Collective, 219–221. Winnipeg: Arp Books.

Kuppers, Petra. 2014. Edges of Water and Land: Transnational Performance Practices in Indigenous/Settler Collaborations. *Journal of Arts & Communities* 6 (1): 5–28.

Kwaymullina, Ambelin. 2018. You Are on Indigenous Land: Ecofeminism, Indigenous Peoples and Land Justice. In *Feminist Ecologies: Changing Environments in the Anthropocene*, ed. Lara Stevens, Peta Tait, and Denise Varney, 193–208. London; New York: Palgrave Macmillan.

Leduc, Timothy B. 2010. *Climate, Culture, Change: Inuit and Western Dialogues with a Warming North.* Ottawa: University of Ottawa Press.

Lewis, Simon L., and Mark A. Maslin. 2015. Defining the Anthropocene. *Nature* 519 (7542): 171–180.

Lewis, Simon L., and Mark Maslin. 2018. *The Human Planet: How We Created the Anthropocene.* London: Pelican.

Looking Horse, Chief Arvol. 2018. Standing Rock Is Everywhere: One Year Later | Chief Arvol Looking Horse. *The Guardian*, February 22, sec. Environment. https://www.theguardian.com/environment/climate-consensus-97-percent/2018/feb/22/standing-rock-is-everywhere-one-year-later.

Lucas, Anne. 2004. No Remedy for the Inuit: Accountability for Environmental Harms under U.S. and International Law. In *New Perspectives on Environmental*

Justice: Gender, Sexuality, and Activism, ed. Rachel Stein, 191–208. New Brunswick: Rutgers University Press.

Malm, Andreas, and Alf Hornborg. 2014. The Geology of Mankind? A Critique of the Anthropocene Narrative. *The Anthropocene Review* 1 (1): 62–69.

May, Theresa J. 2010. Kneading Marie Clements' Burning Vision. *Canadian Theatre Review* 144 (1): 5–12.

———. 2014a. Indigenous Theatre in Global Times: Situated Knowledge and Ecological Communities in Salmon Is Everything and Burning Vision. In *Performance on Behalf of the Environment*, ed. Richard D. Besel and Jnan A. Blau, 193–210. Lanham, Maryland: Lexington Books.

———. 2014b. The Education of an Artist. In *Salmon Is Everything: Community-Based Theatre in the Klamath Watershed*, 103–140. Corvallis: Oregon State University Press.

May, Theresa J., and Klamath Theatre Project. 2014. Salmon Is Everything. In *Salmon Is Everything: Community-Based Theatre in the Klamath Watershed*, 23–92. Corvallis: Oregon State University Press.

McCovey, Kathleen. 2014. I Am Karuk! My Voice as Rose. In *Salmon Is Everything: Community-Based Theatre in the Klamath Watershed*, ed. Theresa J. May, 93–102. Corvallis: Oregon State University Press.

McLean, Sheelah. 2014. Idle No More: Re-Storying Canada. In *The Winter We Danced: Voices from the Past, the Future, and the Idle No More Movement*, ed. The Kino-Nda-Niimi Collective, 92–95. Winnipeg: Arp Books.

McMahon, Ryan. 2014. The Round Dance Revolution: Idle No More. In *The Winter We Danced: Voices from the Past, the Future, and the Idle No More Movement*, ed. The Kino-Nda-Niimi Collective, 98–102. Winnipeg: Arp Books.

Mitchell, Audra. 2015. Decolonising the Anthropocene. *Worldly*, March 17. https://worldlyir.wordpress.com/2015/03/17/decolonising-the-anthropocene/.

Morin, SkyBlue Mary. 2014. A Healing Time. In *The Winter We Danced: Voices from the Past, the Future, and the Idle No More Movement*, ed. The Kino-Nda-Niimi Collective, 7–9. Winnipeg: Arp Books.

Neimanis, Astrida. 2017. *Bodies of Water*. London; New York: Bloomsbury Academic.

Nolan, Yvette. 2014. The Collapse of Worlds in Laura Shamas's Chasing Honey. In *Enacting Nature: Ecocritical Perspectives on Indigenous Performance*, ed. Birgit Däwes and Marc Maufort, 105–115. Brussels: P.I.E. Peter Lang.

———. 2015. *Medicine Shows: Indigenous Performance Culture*. Toronto: Playwrights Canada Press.

Rÿser, Rudolph C. 2015. How Many Indigenous People? *Center for World Indigenous Studies*. Accessed March 03, 2020. https://www.cwis.org/2015/05/how-many-indigenous-people/.

Sinclair, Niigaanwewidam James. 2014. Dancing in a Mall. In *The Winter We Danced: Voices from the Past, the Future, and the Idle No More Movement*, ed. The Kino-Nda-Niimi Collective, 148–150. Winnipeg: Arp Books.

Smith, Linda Tuhiwai. 2012. *Decolonizing Methodologies: Research and Indigenous Peoples*. 2nd ed. London: Zed Books.

Stanlake, Christy. 2009. *Native American Drama: A Critical Perspective*. Cambridge; New York: Cambridge University Press.

Todd, Zoe. 2015. Indigenizing the Anthropocene. In *Art in the Anthropocene: Encounters Among Aesthetics, Politics, Environments and Epistemologies*, ed. Heather Davis and Etienne Turpin, 241–254. London: Open Humanities Press.

———. 2017. Fish, Kin and Hope: Tending to Water Violations in *Amiskwaciwâskahikan* and Treaty Six Territory. *Afterall: A Journal of Art, Context and Enquiry* 43 (March): 102–107.

Tuck, Eve, and K. Wayne Yang. 2012. Decolonization Is Not a Metaphor. *Decolonization: Indigeneity, Education & Society* 1 (1): 1–40.

Vizenor, Gerald. 1999. *Manifest Manners: Narratives on Postindian Survivance*. First Bison Books Printing edition. Lincoln: University of Nebraska Press.

———. 2009. *Native Liberty: Natural Reason and Cultural Survivance*. Lincoln: University of Nebraska Press.

Watt-Cloutier, Sheila. 2015. *The Right to Be Cold: One Woman's Story of Protecting Her Culture, the Arctic and the Whole Planet*. Toronto: Allen Lane.

Woynarski, Lisa. 2015a. Ecological Sentinels: Indigenous Heroes or Colonial Cliché? *Research in Drama Education: The Journal of Applied Theatre and Performance* 20 (2): 186–190.

———. 2015b. Locating an Indigenous Ethos in Ecological Performance. *Performing Ethos: International Journal of Ethics in Theatre & Performance* 5 (1): 17–30.

Woynarski, Lisa, Adelina Ong, Tanja Beer, Stephanie Beaupark, Jonah Winn-Lenetsky, Rulan Tangen, and Michelle Nicholson-Sanz. 2020. Climate Change and the Decolonized Future of Theatre. *Theatre Research International* 45 (2): 179–207.

Yusoff, Kathryn. 2019. *A Billion Black Anthropocenes or None*. Minneapolis: University of Minnesota Press.

CHAPTER 7

Conclusion: Imagining the Future(s)

One evening in early autumn, I followed the multiple staircases down to the depths of the Barbican in London, arriving at the Pit Theatre. I was given a handmade fabric satchel and told to check my bag—the satchel contained everything I would need. I was led through backstage corridors, through storage areas for unused set and props, before finally arriving for my 'shift' at the Factory of the Future. This was the interactive performance work *WE KNOW NOT WHAT WE MAY BE* (2018) by Metis Arts. It set out to discuss, imagine and collaboratively create alternative visions of the future based on equality, inclusivity and climate justice. The title is a quote from Ophelia in Shakespeare's *Hamlet* and the aim was to 'work out who we might be in an alternative future, a future that creates a more just society, and in doing so adverts runaway climate change' (Metis Arts 2018). The work focused on economics, exploring other economic models or narratives (rather than neoliberal capitalism) that reflected priorities of humanity, equality, ecology and inclusivity. As audience-participants we were asked to envision ways of doing things differently in the future, putting in place policies in the near future that will benefit the distant future. We were asked to discuss, debate and imagine more ecological ways of living together.

Over five days, there were hourly start times to the performance with the waves of audiences called generations, each shift beginning with a short talk by an academic, writer, campaigner or activist speaker proposing a change to the current status quo. Speakers included Kate Fletcher on

L. Woynarski, *Ecodramaturgies*, New Dramaturgies,
https://doi.org/10.1007/978-3-030-55853-6_7

fashion, Fran Boait on changing money, Wanda Wyporska on inequality and Paul Mason on social networks for change. Amongst set storage in a large backstage space, I heard campaigner Barb Jacobson lay out a compelling case for universal basic income, or a social dividend, which would help bring gender equality and empower those unable to work, while also reframing a person's value away from 'productivity' in a neoliberal sense, bringing better health and more happiness from having to work less. I was certainly convinced of the need for a social dividend, paid to every adult citizen. After the talk, large doors behind us opened and we entered the Factory of the Future in the transformed Pit Theatre. The work took place on the flat stage area, divided up in different time periods (2020s and 2040s), each with specific actions (a map was included in the programme). We entered in the 2020s, which had cardboard tables and stools set up for conversation, with iPads encased in the tables and paper-covered lights printed with phrases such as 'Act Now' and 'Regenerate to Create' (Fig. 7.1). 'Tragic or epic' was written in large white block letters across

Fig. 7.1 Metis Arts, *WE KNOW NOT WHAT WE MAY BE*, Barbican, London, 2018. (Photo: David Sandison, courtesy of Metis Arts)

the black walls of the theatre, gesturing to the choices we could make about the future. The atmosphere was busy and energised, filled with animated conversation and people moving around the spaces. Recorded bird song would play to prompt us to move around or make commitments to specific future actions. We were invited by performer-facilitators to sit at different tables to discuss near-future scenarios, written on postcards, based on existing, but not yet implemented, propositions. We were each given a series of dot stickers which represented our votes. We could use them to vote different scenarios into the future. The scenarios could also be amended or altered through conversations. These conversations carried on through different waves of generations of audiences. They included proposals such as the right for people to reclaim disused or derelict land, shareholder profits can only be released once a clean supply chain is proved, a real living wage, a citizen's income, a high-rate plastic tax and a life fund (entitling every adult to seven years paid sabbatical).

Each scenario was presented with a research context, a story and questions to stimulate the conversations. These were provocations for thinking differently about how we want to live and be together. Presented as a series of potentials rather than a singular panacea or 'solution', they were intended to be starting points for further conversation, carrying on after the performance. I discussed and debated these scenarios and their possible effects with fellow audience members who were strangers to me with different life experiences and potentially different political and ideological orientations. It was about imagining better futures together, not one singular idea of utopia but collaboratively creating multiple visions for better living for everyone.

As I have argued in Chap. 2, intersectional ecological thinking is needed to understand how climate change effects are distributed along gender, race, class, disability, sexuality and regional lines. The approach needs to be feminist, anti-racist, inclusive, queer-affirming, non-anthropocentric and decolonised. Although based primarily on the UK context, *WE KNOW NOT WHAT WE MAY BE* addressed the intersectional impact of the future scenarios. For example, when a carbon tax is introduced, it is with a condition of price controls to protect those most vulnerable. The move to renewable energy also has impacts beyond the UK when implemented in the future:

> Some multinationals have seen opportunities in desert solar power. This isn't such good news for those living around the edges of the Sahara desert. The people in Mauritania and Morocco haven't seen much benefit from the hundreds of miles of solar fields covering their lands, given much of the power is cabled to Europe, or charged back to them at extortionate prices. There are regular attacks on the fields, with protesters calling for the World Bank to make it mandatory for energy to be publicly owned to be eligible for a world bank loan. (Metis Arts n.d.)

By calling attention to the unintended effects of renewable energy transitions, the participants were prompted to think about and discuss the complexities and multiple impacts across borders of actions taken in the UK. They were asked the intersectional ecological questions of who and what is represented or excluded? What ideology, knowledge or subject position is represented as the 'norm' or dominant? In what ways are ecological issues also inherently tied to social, political, gendered, racialised, class-based, disability and access issues?

The bioperformativity of assemblages like the power grid, as discussed in Chap. 3, is also evident in the scenario of renewable energies. The new structures of renewables have agency, not only in creating new 'clean' energy sources for Europe but also for the impact on the people of Mauritania and Morocco. Every part of the assemblage performs in a way that creates socio-ecological and economic effects. In some ways, the binaries between human/nature and urban/nature were made malleable in the scenarios, where innovative urban food growing techniques were modelled and green roofs and living walls were imagined as covering the majority of city buildings. However, there was an anthropocentric dominance to much of the future visioning, centring humans and often bracketing off non-humans. Only one scenario addressed non-human wildlife, which was the creation of protected climate corridors across former motorways (no longer widely used due to improvements in public transportation) for animals to move habitats as the climate continues to change. Instead of making neighbourhoods and cities places where more-than-humans can thrive, they were sequestered to specific human-created boundaries. Anthropocentric as the dominant representational mode remained intact.

After leaving the 2020s area of the performance, I passed the Flyway to the Future where the scenarios that had been voted into the future were displayed. Then I entered the 2040s where the details of the future or the 'new normal' were imagined, having implemented the scenarios voted through. There was a large cardboard model of a city and hydroponically growing salad. There was also a Collaboratory where a series of short performances imagined what people in the 2040s might be thinking about. These included a radio show that aimed to 'look back to the crazy teens of the twenty-first century' (Metis Arts 2018), remakes of disaster movies from the past imagined in the alternative future, and a city tour. Headphones in the raked seating banks were offered for audience viewing and listening to these performances. Possible ideas of alternative futures were modelled and enacted through these mini-performances, allowing the audience/participants to envision how the possible futures scenarios they voted through would play out.

The work was immersive and participatory in the decision-making process as well as the modelling in the Collaboratory. As I argued in Chap. 4, immersion and participation in ecological contexts can reveal the way in which ecological relationships are differently embodied. Vulnerability and exposure of bodies to climate change effects characterise this embodiment. In the future scenario, the increase in vegan and local diets means 'most people now refuse to eat food that has been "doped", so the market for pesticides has collapsed' (Metis Arts n.d.), addressing one of the ways in which different bodies are differentially exposed to toxins and ecological impacts of the neoliberal, colonialist ideology. Care is also accounted for in these future scenarios, where new large-scale compassionate care centres and Citizen's Income have empowered the elderly and those with disabilities that require care. Some differences are accounted for in the future scenarios, although with the reduction in plastic production I did not see any mention of the impact this would have on medical tools or accessibility for people with disabilities (viable alternatives for those that require plastic bendy straws to drink, for example). Although class and age were considered explicitly, to a certain extent, there was still a presumptive normative body for whom these scenarios were enacted for. A further iteration of the work, *Factory of the Future*, for the Oslo Architecture Triennale in 2019 addressed these issues more explicitly.

Environmental justice issues were addressed by scenarios that provided housing and green spaces (green roofs, living walls) so that all people, regardless of race or class could access clean air and an affirmative 'homeplace' (hooks 1990), as I describe in Chap. 5. A law was passed that ownership of all land and property must be registered to a named, accountable individual, making offshore ownership of land and housing impossible.

There are also no more landlords and people are not able to own more than one residential property, driving house prices in London down. There are intergenerational residences and more people decided to live together. The gender and racial pay gaps close due to the Citizen's Income, the Real Living Wage and the Life Fund which increases access to universities so that the gap between access to education, background and social mobility closes. The idea of home shifts, to one not centred on private ownership and consumer accumulation, but to a community-based ideal where local pubs are community-owned businesses, public spaces are accessible, and neighbourhoods control and administer energy as well as health care and street parties of cook-ups from supermarket food waste are common (as a tax is introduced on supermarket food waste). There seems to be more time and investments in arts and pleasurable activities as there is less consumption and more community. The ideas voted through represent de-centralised, small-scale policies that allow for a more tailored approach to specific needs of a community, determined by a citizen forum, rather than a postcode lottery, or income being a determining factor in the quality of air where you live. The focus on collectivity has the potential to break some of the structures that continue to profit from climate crisis.

The local focus of some of the scenarios does not mean a narrowed idea of global ecological relationships. In my reading, the piece was able to reframe the colonialist narrative of the Anthropocene, which I articulated in Chap. 6, by addressing structural change, stripping the power of private corporations and the governmental systems that enable them to pollute unencumbered, exploiting workers and the more-than-human world. It was not humans writ large that needed to change but the power structures that govern our lives through economics. Within the collective vision of the piece, a clean supply chain and Trace-it Tax means that any goods purchased draw attention to the contexts, labour and resources that produced them. 'The TRACE-IT TAX [which requires clear labelling of the supply chain and manufacturing processes] has had a significant impact in China, Vietnam, Bangladesh. Being paid a better wage and experiencing less environmental degradation has transformed the lives of millions of textile workers' (Metis Arts n.d.). The potential for an eco-cosmopolitan shift in thinking about ecological relationships, from Chap. 5, is present in this modelling. When you purchase an item of clothing, where it came from, how it was made and who made it would be clearly labelled with an extra cost applied via a tax for harmful materials, production practices and pollution generated. In practice this could transform the everyday relationships we, in the Global North, have with consumer goods, recognising the way in which we are connected to workers, materials and pollution in distant places.

Will *WE KNOW NOT WHAT WE MAY BE* lead directly to an alternative, more just future? No, probably not. But the sensation or affect of the performance lingered beyond the experience (Thompson 2009: 178). It was exciting to imagine ecologically just futures and debate the scenarios and policies that can facilitate it, mindful of the way in which decisions based on a British context impacted those living in other places, and those without the political or social capital to contribute to the debates. I left the performance thinking that perhaps these potential futures are not so far away and made a stronger commitment to voting for policies that reflect these ideas. I also felt anger and frustration as to why many of these common sense, good-for-people-and-environment scenarios have not yet been implemented and why those who profit from the current status quo, and therefore have little motivation to change it, currently govern us. The question of how the theatrical imagining of the performance can translate into political action remains.

In many ways we might think of the current climate crisis as a failure; a failure of action and a failure to stop colonialist and neoliberal power structures that legally prioritise profit over people and the more-than-human world. We might also think of theatre and performance as complicit in this failure. This brings up the well-rehearsed debates about what theatre and performance can do. Chaudhuri and Enelow assert that theatre fails in terms of representing climate change as it has to be engaged with cognitively rather than phenomenologically, because 'the only way it can be apprehended is through data and modelling—through systems and mediations—all of which have to be processed cognitively and intellectually: have to, in short, be understood, rather than experienced, phenomenologically and temporally' (2014: 23). Throughout this book, I argue that this is not a complete failure because, although climate change may not be able to be experienced as a whole, different aspects of our relationship to the more-than-human and the current ecological situation (including who is excluded or exposed through differentiated power structures) can be experienced and made manifest in theatre and performance. According to Tony Fisher, it is 'theatre's *failure* to "represent" the world that connects it to the world' (2019: 180 original emphasis). Fisher writes that the ethical and political task of theatre is 'to think the impossible possibility' of the 'reworlding of the world' (188). It is perhaps then the 'reworlding' through dramaturgies of revealing (neo)colonial logics, interconnections, weathering, resistance and activism that is what theatre and performance can do.

As I articulated in the Introduction, I do not think theatre and performance are a blunt instrument for social change: they will not be able to change behaviour or power structures in direct correlation. They can, however, help us understand our failure better. Failure can open up new narrative potentials or be a mode of resistance, as failure is tied to capitalism as Jack Halberstam notes: 'a market economy must have winners and losers, gamblers and risk takers, con men and dupes' where success is related to profit (Halberstam 2011: 88). Therefore, failure can represent a mode of living, an alternative to neoliberal constructs of success, towards collectivity instead. When theatre and performance can provide a space for that resistance, failure can counter hegemonic narratives against the 'US imperialist project of hope' (Halberstam 2011: 106) and the belief that governments and/or technology will prevail in averting ecological crisis. Ecodramaturgies can also acknowledge limitations, 'about the narrowness of the future, the weightiness of the past, and the urgency of the present' (Halberstam 2011: 106) in our current ecological situation. Both Halberstam and José Esteban Muñoz understand queer failure as a liberatory concept: 'within failure we can locate a kernel of potentiality' (Muñoz 2009: 173). Failure is rejection of 'normative ideas of value' (Muñoz 2009: 173), and in this way becomes a mode of resistance. Within the context of this book, I find resonance with the idea of the failure of systems and structures to address climate emergency as the failure of the heteropatriarchal, colonialist logic of those systems and structures—these are failing us (humans and more-than-humans) in numerous ways. What *WE KNOW NOT WHAT WE MAY BE* shows us is that we must fail as 'good neoliberal subjects' in order to envision a way of living together that values equality, inclusion and climate justice. Ecodramaturgies provides a process of doing that, but what happens after and how does the experience relate to the world at large? This may be the failure.

My thinking about ecodramaturgies has been offered here as a suggestion of potential ways of critiquing and interpreting theatre and performance practices within an ecological context, rather than a descriptive set of strategies. The broad spectrum of practices included here are works that create a space for dialogue, and/or provide imaginative possibilities for ways of (re)thinking and relating to the world (in different ways and to varying degrees of profundity). Perhaps then, as Heddon and Mackey suggest, affect is produced through the relationship between reality and aesthetic practices (2012: 176). Theatre and performance facilitate different modes of engagement, eliciting imaginative, creative and affective responses. In this book, I have argued that ecodramaturgies are created by performance modes of engagement that critique ecological relationships by making manifest their effects (such as pesticide poisoning, contribution to melting ice caps, crazy weather, rising temperatures and hungry polar bears) and the intimate interconnections to our everyday lives. By reframing some of the pervasive and reductive images and tropes of ecology, ecodramaturgical practices and modes of meaning-making raise questions about the relationship between humans and the more-than-human world, reframing exclusionary narratives of the 'human' using dramaturgies of meshwork and interrelations.

As I finish editing this book, in April 2020, the UK has entered lockdown in response to the global pandemic of the novel coronavirus (COVID-19). In some ways, the pandemic has demonstrated how connected and interrelated we are—biologically to each other and the world around us, to global trade systems and politicalised structures of healthcare. Responses to the pandemic (from income support to mutual aid groups) have shown that governments can react quickly and find money when they need to. The climate emergency demands a similar response, which we now know is possible. The virus is following established patterns of inequality affecting vulnerable people, people of colour, Indigenous peoples, economically precarious people, refugees and prisoners disproportionately as Schifferes notes: 'within cities it is the poorer areas that have the highest infection rates. In London, for example, poorer inner boroughs such as Lambeth and Southwark have triple the per-capita infection rates than the leafy suburbs of Kingston, Sutton and Bexley' (Schifferes 2020). The situation is intensified by overcrowded social housing and working class jobs that cannot be done from home and require travel. In the US, African-Americans seem more likely to die from the virus than any other group (Taylor 2020). Key workers are revealed to be the people who are among the lowest paid in society: carers, nurses, grocery store workers, bus drivers, delivery people, postal workers. The tension between protecting people and 'restarting the economy' continues to be a deeply politicised one in many countries. There are some positive signs as well: in the UK for example, air pollution was down by 60% compared to the previous year (Monks 2020) and there has been a historic run of coal-free days in power generation (Ambrose and Kommenda 2020). Spain has introduced a universal basic income. There are strong voices demanding that post-pandemic we do not return to business as usual. This may be an opportunity to rebuild towards a low-carbon, just society based on equality, access and social solidarity. At the time of writing, the situation is still ongoing, and the full effects of the pandemic are not yet known.

Despite the climate emergency and the failure to act on it, I continue to focus my work on these seemingly impossible problems, taking solace in the moments of resonance, reflection, thoughtfulness, pleasure, fear and anger found in the experiences of theatre and performance. Ecodramaturgies can open up potential spaces to share, confront and dissect anxieties and uncertainties about the complex socio-ecological conditions we live in.

References

Ambrose, Jillian, and Niko Kommenda. 2020. Britain Breaks Record for Coal-Free Power Generation. *The Guardian*, April 28, sec. Energy Industry. https://www.theguardian.com/business/2020/apr/28/britain-breaks-record-for-coal-free-power-generation.

Enelow, Shonni, and Una Chaudhuri. 2014. Theorizing Ecocide: The Theatre of Eco-Cruelty. In *Research Theatre, Climate Change, and the Ecocide Project*, ed. Una Chaudhuri and Shonni Enelow, 22–40. London; New York: Palgrave Macmillan.

Fisher, Tony. 2019. Theatre of the Worldless. In *Beyond Failure: New Essays on the Cultural History of Failure in Theatre and Performance*, ed. Tony Fisher and Eve Katsouraki, 169–191. London; New York: Routledge.

Halberstam, Jack. 2011. *The Queer Art of Failure*. Durham: Duke University Press Books.

Heddon, Deirdre, and Sally Mackey. 2012. Environmentalism, Performance and Applications: Uncertainties and Emancipations. *Research in Drama Education: The Journal of Applied Theatre and Performance* 17 (2): 163–192.

hooks, bell. 1990. *Yearning: Race, Gender, and Cultural Politics*. Boston: South End Press.

Metis Arts. 2018. *WE KNOW NOT WHAT WE MAY BE Programme*. Barbican, London, September 6–9.

———. n.d. Emergent Scenarios. *Metis Arts*. Accessed May 20, 2019. https://metisarts.co.uk/projects/we-know-not-what-we-may-be/the-story/emergent-scenarios.

Monks, Paul. 2020. Coronavirus: Lockdown's Effect on Air Pollution Provides Rare Glimpse of Low-Carbon Future. *The Conversation*, April 15. https://theconversation.com/coronavirus-lockdowns-effect-on-air-pollution-provides-rare-glimpse-of-low-carbon-future-134685.

Muñoz, José Esteban. 2009. *Cruising Utopia: The Then and There of Queer Futurity*. New York: New York University Press.

Schifferes, Steve. 2020. The Coronavirus Pandemic Is Already Increasing Inequality. *The Conversation*, April 10. https://theconversation.com/the-coronavirus-pandemic-is-already-increasing-inequality-135992.

Taylor, Keeanga-Yamahtta. 2020. The Black Plague. *The New Yorker*, April 16. https://www.newyorker.com/news/our-columnists/the-black-plague.

Thompson, James. 2009. *Performance Affects: Applied Theatre and the End of Effect*. London; New York: Palgrave Macmillan.

Index[1]

[1] Note: Page numbers followed by 'n' refer to notes.

L. Woynarski, *Ecodramaturgies*, New Dramaturgies,
https://doi.org/10.1007/978-3-030-55853-6

B

Q

R

The manufacturer's authorised representative in the EU is Springer
Nature Customer Service Centre GmbH, Europaplatz 3, 69115 Heidelberg,
Germany. If you have any concerns regarding our products, please
contact ProductSafety@springernature.com

Printed and bound by CPI Group (UK) Ltd, Croydon, CR0 4YY
30/09/2026
02225460-0003